The Permian Extinction and the Tethys: An Exercise in Global Geology

A.M. Celâl Şengör
İstanbul Teknik Üniversitesi
Avrasya Yerbilimleri Enstitüsü
Ayazağa 34469
İstanbul, Turkey

Saniye Atayman
İstanbul Teknik Üniversitesi
Avrasya Yerbilimleri Enstitüsü
Ayazağa 34469
İstanbul, Turkey

THE
GEOLOGICAL
SOCIETY
OF AMERICA®

Special Paper 448

3300 Penrose Place, P.O. Box 9140 ▪ Boulder, Colorado 80301-9140, USA

2009

Published by The Geological Society of America, Inc.
3300 Penrose Place, P.O. Box 9140, Boulder, Colorado 80301-9140, USA
www.geosociety.org

Printed in U.S.A.

GSA Books Science Editors: Marion E. Bickford and Donald I. Siegel

Library of Congress Cataloging-in-Publication Data

Şengör, A. M. Celâl.

The Permian extinction and the Tethys : an exercise in global geology / A.M. Celâl Şengör and Saniye Atayman.
   p. cm. — (Special paper ; 448)
  Includes bibliographical references and index.
  ISBN 978-0-8137-2448-5 (pbk.)
   1. Geology, Stratigraphic—Permian. 2. Tethys (Paleogeography) 3. Extinction (Biology) 4. Anoxic zones. I. Atayman, Saniye. II. Title.

QE674.S42 2009
551.7′56—dc22

2008046587

Cover: The tempestuous seaside scene is a woodcut from R. Bommeli's *Die Geschichte der Erde* (*The History of the Earth*) (1890, p. 335) and is there entitled "ideal landscape of the Dyassic [Permian] time showing a volcanic eruption." Bommeli reminded his readers that most geologists in those days were in agreement about the overall poverty of the Permian fauna. He pointed out that some had sought to explain this "on the basis of the occurrence of great masses of eruptive rock, especially of porphyry, ... that the end of the Palaeozoic era must have been a very 'stormy' time, during which volcanic revolutions had altered the relationships of land and sea and especially damaged the fauna and the flora." He then says, "But against this view so many apposite objections have been put forward that it had to be abandoned" (Bommeli, 1890, p. 342). In his book, Bommeli prefers Eduard Suess's student and later colleague Theodor Fuchs's idea of a more local phenomenon resulting from the restriction of certain sea basins, such as that of the Zechstein in northern Germany, and the resulting chemical changes of the seawater (Bommeli, 1890, p. 350). Because we think submarine eruptions may well have triggered gas eruptions in a restricted Paleo-Tethys, we chose Bommeli's wonderful woodcut of a submarine eruption causing commotion in a stormy and foggy seascape, possibly already with a poisonous atmosphere!

The skull belongs to the *Dicynodon lacerticeps* Owen, described and figured by Sir Richard Owen first in 1845 and then again in the first edition of his *Palæontology* (1860, p. 237). The figure here was copied from his *Palæontology*. It stands for all Dycinodont reptiles that used the Cathaysian bridge to cross from Gondwana-Land to Laurasia and thus now help us document the former existence of that bridge that almost sealed the Palaeo-Tethys against the Panthalassa in the late Permian world.

Cover design by Heather L. Sutphin

10 9 8 7 6 5 4 3 2 1

Dedicated to William Richard Dickinson,
geologist, scholar, teacher, gentleman, and friend.

What is the point of writing only about those that we see?
Why don't we write a little about those that we think?

—*Abdülhak Hâmit Ṭarhan,
late nineteenth or early
twentieth century*

Hypotheses are nets; only those who cast will catch.
Was not America itself found by hypothesis?
Above all hail to the hypothesis—only it remains
Forever new, as long as it can conquer itself.

—*Novalis, 1798*

Ideas without precedent are generally looked upon with
disfavor and men are shocked if their conceptions of an
orderly world are challenged.

—*J. Harlen Bretz, 1928*

This was the period of the Zechstein with its impoverished
marine fauna...We have now reached one of the most
remarkable phases of the earth's history.

—*Eduard Suess, 1888*

Contents

Preface and Acknowledgments

This book grew out of an invitation to present a paper in the William R. Dickinson Symposium, entitled Circum-Pacific Tectonics, Geologic Evolution, and Ore Deposits, held in Tucson, Arizona, from 24–30 September 2007. Both of us enthusiastically accepted the invitation in view of the great honor thus bestowed upon us to speak in a symposium lauding one of the greatest geologists of our times.

However, it was not easy to prepare a paper that would be worthy of the laureate. Apart from their abundance, three significant aspects have so far characterized Bill Dickinson's contributions to human knowledge: (1) originality of concepts, (2) novelty and accuracy of observations to test concepts, and (3) emphasis on relations of new concepts to old ones to show where improved understanding lies. When invited to contribute a paper to the Dickinson Symposium, we were determined to keep Bill's spirit and consequently have sought what we believe to be a new topic, presented with hopefully accurate observations, which to us appear to represent a solution to an old problem. The frame of our topic had been set by Jon Spencer, one of the organizers, in his invitation, as the "Tethys," and the problem we chose within that frame is the cause of the so-called "end-Permian" extinction and its relation to the Tethyan realm. What led us to that particular problem was the growing recognition that the late Permian extinction had affected organisms living in the sea as well as on land and had been most likely related somehow to diminished oxygen in both environments. Oxygen diminution seemed to have started in a big way in a current-restricted, or even continent-locked, Paleo-Tethys, to which and to its surroundings, most of the extinction events appear confined, or at least in and around which had been most intense.

Having decided that our topic was as grand as we could handle, the next difficulty was to do justice to it. Merely attempting to do so took a good part of a whole year of literally round the clock work by both of us. We cannot tell whether we succeeded in our goal of at least pointing the way to a viable solution to the problem we chose to tackle—this will be the decision of our critics—but we definitely did fail to comply with the space restrictions that had been set in the original invitation. As Ben Franklin is reputed to have said once, our script would have been shorter had we had more time to write it.

When Jon Spencer received our paper, he did not immediately reject it out of hand owing to its incommensurate size; instead, he carefully reviewed it himself, encouraged us to find other competent reviewers and send the results to him, and he finally said that while the topic was appropriate and its handling acceptable, he had failed to find a way to include the paper into the proceedings volume of the Dickinson Symposium on account of its size. He asked us to prepare a summary within the limits set by the invitation guidelines, perhaps to put some emphasis on ore genesis in view of the symposium theme and the large volumes of black shales we discuss, and to send it to him. However, he also strongly urged us to think about publishing our original manuscript in book form and suggested we submit it to the consideration of the Geological Society of America's Books Science Editor Professor Donald Siegel with a view to possible publication as a GSA book. Professor Siegel promptly replied and invited us to send him our manuscript. Following one of the fastest and most helpful reviewing episodes we have ever been blessed with as authors, he accepted our book to be published as a Geological Society of America Special Paper.

A study of this scope owes so much to so many that it is hardly possible to squeeze them all into a simple statement of acknowledgement. First of all, let it be known that if it were not for Jon Spencer and Donald Siegel this book would have never seen the light of day. We are immensely grateful to them both for their help and guidance. Our gratitude to Bill Dickinson is impossible to put into mere words: first and foremost he has provided—and, happily, still does so—such a formidable example to us both in addition to countless others he trained and/or influenced. It was under the perennial terror of not being able to produce something worthy of him that we have labored day and night—although we are still not sure whether our labor has actually brought us to the point where we can confidently lay its results at Bill's feet. Bill also deserves more than a simple note of gratitude for teaching the senior author, among other things, so much geology over a lifetime in an atmosphere of almost parental affection. The junior author, a student of the senior author, is thus an indirect beneficiary of Bill's teaching and is accordingly grateful.

Bernard Battail, Bob Berner, Greg Davis, Bob Diaz, Maurizio Gaetani, Bob Garrison, Niyazi Gündoğdu, Yukio Isozaki, Conrad Labandeira, Spencer Lucas, Jean Marcoux,[1] Boris Natal'in, Sinan Özeren, David Rowley, Bruce Rubidge, Şevket Şen, Chris Scotese, Shen Shu Zhong, Rudolf Trümpy, and Xiao Wenjiao all helped us by informing us of developments in their respective fields or in their institutions, and all of those whom we asked allowed us to use their unpublished data and many sent us publications at very short notice. Fabrizio Cecca, of the University of Paris-6 (Pierre et Marie Curie), invited us to present our model in the first international symposium on paleobiogeography (*1er Symposium International de Paléobiogégraphie*, 10–13 July 2007, Paris, France) and our scholarly friend, the great vertebrate paleontologist Philippe Janvier, who was sitting in the audience, insisted immediately after our presentation that we publish it. Dan McKenzie, as always, allowed us to tap his prodigious brain for useful advice.

We have been immensely lucky with our reviewers: Demir Altıner, Ankara; Robert E. Garrison, California; Spencer G. Lucas, New Mexico; and Sébastien Steyer, Paris, who have helped us more than is usual with reviewers. They not only pointed out our errors and omissions, but also supplied us with advice and literature to enable us to correct and to complete our manuscript. We absolve, however, all of the named from any responsibility for our outrageous statements and possible consequent errors.

Mr. Ayhan Kaygusuz, the great head librarian of the Istanbul Technical University, performed miracles in getting us obscure publications in record time. We are very much indebted to Professor Okan Tüysüz, the director of the Eurasian Institute of Earth Sciences; to Professor Mahir Vardar, the dean of the Faculty of Mines; and to Professor Faruk Karadoğan, the *rector magnificus* of the İstanbul Technical University, for allowing us to break nearly all university regulations under the sun to enable us to keep to our frantic schedule. Professor Karadoğan also went out of his way to create funds in a last minute rush to enable a negligent Saniye to go to the Geological Society of America Annual Meeting in 2007 in Denver to present a part of our work before noted experts in the various fields.

Mrs. Oya Şengör deserves warm thanks for looking after Saniye and providing efficient shielding against Celâl's fits of temper while Saniye stayed with the Şengörs for long stretches of time as this book was being written. Mr. Berkin Atayman sacrificed much time and effort in helping us with our figures and generously agreed to spend what amounted to a lonely and celibate year for the cause of geology. Finally Mr. Asım Şengör is thanked for his help in chemistry and for regularly annoying Saniye into repeated bursts of furious creative activity.

A.M. Celâl Şengör, İstanbul
Saniye Atayman, Johannesburg
20 June 2008

---

[1]While the typescript of this book was under revision, the terrible news of Jean Marcoux's death arrived (17 June 2008). Since the sixties, Marcoux has made, among many other things, immense contributions to the Permian-Triassic boundary problems in the Tethyan realm. His disappearance has dealt a severe blow to Tethyan studies.

The Geological Society of America
Special Paper 448
2009

The Permian Extinction and the Tethys:
An Exercise in Global Geology

A.M. Celâl Şengör
Saniye Atayman*
Istanbul Teknik Üniversitesi, Avrasya Yerbilimleri Enstitüsü, Ayazağa 34469, Istanbul, Turkey

ABSTRACT

The Tethyan realm stretches across the Old World from the Atlantic to the Pacific Oceans along the Alpine-Himalayan mountain ranges and extends into their fore- and hinterlands as far as the old continental margins of the now-vanished Tethyan oceans reached. It contains the Tethyside superorogenic complex, including the orogenic complexes of the Cimmerides and the Alpides, the products of the closure of the Paleo- and the Neo-Tethyan oceans, respectively. Paleo-Tethys was the oceanic realm that originated when the late Paleozoic Pangea was assembled by the final Uralide–Scythide–Hercynide–Great-Appalachide collisions. It was a composite ocean, i.e., not one formed by the rifting of its opposing margins, and its floor was already being consumed along both Laurasia- and Gondwana-Land–flanking subduction zones when it first appeared. The Gondwana-Land-flanking subduction systems, in particular, created mostly extensional arc families that successively led to various Paleo-Tethyan marginal basins, the last group of which was the oceans that united to form the Neo-Tethys.

The Paleo-Tethys may have become an entirely continent-locked ocean through the construction, to the east of it, of a Cathaysian bridge uniting various elements of China and Indochina into an isthmian link between Laurasia and Gondwana-Land during the latest Permian, inhibiting any deep-sea connection between the Paleo-Tethys and the Panthalassa. That land bridge may have been responsible for the peculiarities in the distribution of the latest Permian-early Triassic Dicynodonts and possibly some brachiopods, benthic marine microorganisms, and land plants.

The existence of the Cathaysian bridge seems to have helped the formation of anoxic conditions in the Paleo-Tethys. In fact, it seems that the major Permian extinctions began in the Paleo-Tethys and were really mainly felt in it and in areas influenced by it. This isolated setting of the Paleo-Tethys we refer to as a Ptolemaic condition, in reference to the isolated oceans Claudius Ptolemy had depicted on a geocratic Earth in his world map in the second century AD. Ptolemaic conditions are not uncommon in the history of Earth. Today, such a condition is represented by the Mediterranean and its smaller dependencies such as the Black Sea and the South Caspian Ocean. Para-Tethys in the Neogene had a similar but even more isolated setting.

As we see in all these late Cenozoic cases, such Ptolemaic oceans have a major influence on the evolution of the biosphere. The Paleo-Tethys seems to have had a much larger impact than any of its successors owing to its immense size and may have been the key player in the so-called "end-Permian" extinction, which, in reality, was a mid to late Permian affair, with some late phases even in the earliest Triassic. The

*Present address: Bernard Price Institute for Palaeontological Research, University of the Witwatersrand, Private Bag 3, Wits 2050, South Africa; Saniye.Atayman@students.wits.ac.za.

Şengör, A.M.C., and Atayman, S., 2009, The Permian Extinction and the Tethys: An Exercise in Global Geology: Geological Society of America Special Paper 448, 85 p., doi: 10.1130/2009.2448. For permission to copy, contact editing@geosociety.org. ©2009 The Geological Society of America. All rights reserved.

1

Permian extinction happened in at least two main phases, one in the Guadalupian and the other near the end of the Lopingian, and in each phase different animal and plant groups became extinct diachronously, phasing out according to the degree they were influenced by the developing anoxia within the Paleo-Tethys.

What these conclusions suggest is that when investigating the causes of past events, regional geology must always form the foundation of all other considerations. Many speculations concerning the Permian extinction events cannot be adequately assessed without placing their implications into the geography of the times to which they are relevant. A purely "process-orientated" research that downplays or ignores regional geology and attempts to ape physics and chemistry, as is now prevalent in the United States and in western Europe and regrettably encouraged by the funding organizations, is doomed to failure.

Keywords: Paleo-Tethys, Permian, extinction, anoxia, Cathaysian bridge, Laurasia, Gondwana-Land, Cimmerian continent.

CHAPTER I

Introduction

This book embodies the results of an exercise in regional geology with a view to understanding the causes of the so-called "end Permian" extinction, which led to the lowest biodiversity during all of the Phanerozoic and has been known about since the middle of the nineteenth century. What led us to a reconsideration of this particular problem within the framework of regional geology was the growing recognition that the late Permian extinction had been most likely related somehow to low oxygen both in the oceans and in the atmosphere (Graham et al., 1995; Hallam and Wignall, 1997; Berner, 2004, 2007; Hallam, 2004; Huey and Ward, 2005; Ward, 2006; Ward and Berner, 2007; Berner et al., 2007; Twitchett, 2007) and that oxygen diminution in a big way may have started in a current-restricted (Brandner, 1987), or even continent-locked, Paleo-Tethys (Kiehl and Shields, 2005). The purpose of this book is to place these theses in a more radical, and therefore easier to disprove, hypothesis as to their causes and consequences.

In this book we use Gondwana-Land when referring to the supercontinent, as Eduard Suess originally named it (Suess, 1885, p. 768), and Gondwana when referring to the eponymous historical district in India.[2] Contrary to the recommendations of the *International Stratigraphic Guide* (Salvador, 1994), we do not capitalize epoch adjectives, such as late and early, because time cannot be formalized, as opposed to rock, which is a definite object occupying a certain volume. The idea that we can formalize, in a globally valid way, the time interval that elapsed during the formation of a rock package is an illusion resulting from confusing a sort of comprehensive homotaxis with synchrony that has caused untold confusion, from putative "orogenic phases" to "simultaneous global extinctions." The objections Callomon and Donovan (1966), Drooger (1974), and Holland (1989) made to Hollis Hedberg's insistence that chronostratigraphic terms be formalized still apply, after more than four decades of technical advances of measuring geological time. We also use "medial" when referring to time and "Middle" when referring to rock. When referring to figures in the present book, we write Fig. with a capital F; when referring to figures of others that we quote from the literature, we write fig. with a lower case f.

---

[2]It is unfortunate that in the recent geological literature it has become customary to use the name Gondwana for the supercontinent. This is not only wrong, but also confusing, because when referring to the various Gondwana-based entities (flora, plateau, series, basins) it is the district that is meant and not the continent (see Şengör, 1983, 1991a; Rigby and Shah, 1998; and Veevers, 2004). We hyphenate the name, because that was how Suess wrote it and that was how it was introduced into English in the English edition of *Das Antlitz der Erde*.

CHAPTER II

Tethys and the Tethyan realm

The term Tethys was introduced into geology by Eduard Suess (1831–1914) in 1893, when he was trying to show that oceans were not permanent features of the face of Earth.[3] He had known since 1862 that a Triassic seaway had extended from the southern Alps to the Himalaya on the basis of such marine fossils as *Ammonites floridus* Wulf. and *Halobia lommeli* Wissm.,

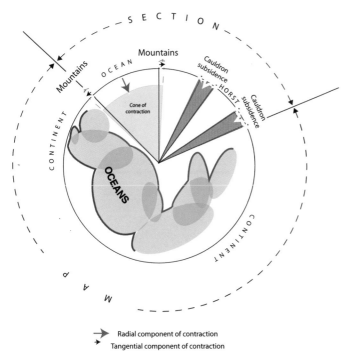

Radial component of contraction
Tangential component of contraction

Figure 1. Eduard Suess's model of terrestrial tectonism, reconstructed from his descriptions, especially his two fundamental publications, *Die Entstehung der Alpen* (*The Origin of the Alps*; Suess, 1875) and *Das Antlitz der Erde* (*The Face of the Earth*; Suess, 1883–1909). Suess divided the effects of the thermal contraction of Earth into a radial component and a tangential component. The radial component was supposed to be expressed by cauldron subsidences, essentially cone-shaped volumes of Earth with apices in the center of the planet. As these sectors contracted in unequal amounts, their bases on the surface of the planet subsided differentially in shapes approximating irregular ellipses. When many such adjacent elliptical areas subsided, their coalescence formed ocean basins. In the figure, the "section" shows cross-sections of contracting volumes. The "map" illustrates how intersecting elliptical subsidences may form oceans. We drew them in such a way as to represent the southern and central Atlantic and the Indian Oceans, although Suess never illustrated such specific examples. His database was simply insufficient. Our figure has the sole purpose of making his theory intelligible to the reader and to show that it was at the time a plausible idea to entertain.

reported abundantly from both places in black, clayey limestones (Suess, 1862, p. 258). By 1875, he had discovered that this seaway had oceanic depths in the Alps (Suess, 1875, p. 101), in places exceeding 4000 m (Neumayr, 1885, p. 137;[4] Suess, 1909, p. 646). Finally, in 1893, on the basis of the observations of his colleagues in the Sillagong region of the Central Himalaya (cf. Diener, 1895, especially p. 551–555 or 19–22 of the offprint), he stated that "Modern geology permits us to follow the first outlines of the history of a great ocean which once stretched across part of Eurasia. The folded and crumpled deposits of this ocean stand forth to heaven in Thibet [*sic!*], Himalaya, and the Alps. This ocean we designate by the name 'Tethys,' after the sister and consort of Oceanus. The latest successor of the Tethyan sea is the present Mediterranean" (Suess, 1893, p. 183).

Suess thought this seaway to have been an ocean with a structure similar to that of the present-day Atlantic (Suess, 1888, p. 681). He believed oceans to have formed by intersecting elliptical subsidences (Suess, 1888, p. 681; see Figs. 1 and 2 herein) and gave the Neogene to Recent history of the Mediterranean as an example of how oceans are made (Suess, 1909, p. 722) and the phenomena of faulting and vulcanicity in Iceland as an example of how the building of the Atlantic was continuing in our own day (Suess, 1888, p. 681).

In Suess's world, oceans were destroyed by shortening. He cited only two examples of "dead" seas not destroyed by tectonic events: the dead sea of the El Djouf desert in the western Sahara[5] and the "dying" Caspian Sea[6] (Suess, 1909, p. 747). He was among the first to appreciate the importance of horizontal motions in mountain building (Suess, 1875, p. 25), and in 1909 he pointed out that the magnitude of such motions was so large as to invite suspicion whether thermal contraction alone could be sufficient to cause them (Suess, 1909, p. 721). The implication was clear that originally raceme-shaped map views of the oceans (cf. Figs. 1 and 2) were transformed into the linear and/or arcuate mountain belts (Figs. 3 and 4) by the subsequent intense shortening and that *the final map views of mountain belts had little relation to the original outlines of the oceans out of which they had grown* (cf. Suess, 1888, p. 681). Suess emphatically rejected the theory of geosynclines, the alleged mother troughs

---

[3]For a controversy as to the paleogeographical meaning Suess attached to the term Tethys, see Tozer (1989, 1990) and Şengör (1990a). For a comprehensive history of the Tethys concept, see Şengör (1998). Also see Kollmann (1992).

[4]Melchior Neumayr (1845–1890), one of the greatest of the Viennese giants in geology and "the greatest paleontologist of his time" according to Viktor Uhlig's famous assessment, was Suess's son-in-law and his colleague at the University of Vienna.

[5]Coincident with the Tindouf Basin.

[6]A small ocean basin, possibly of late Cretaceous age.

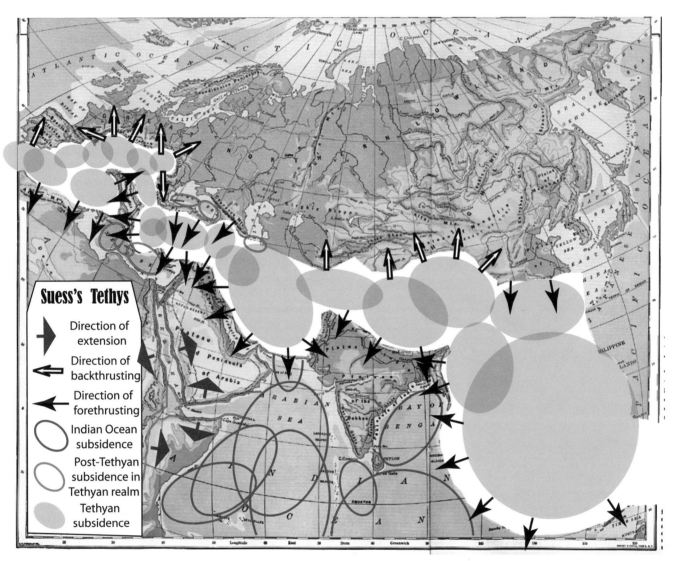

Figure 2. A structure of coalescing elliptical subsidences, as illustrated here, is what Suess may have imagined the Tethys to have had. After having described in detail how the present-day Mediterranean formed by coalescing subsidences, he stated that "The manner in which sea basins arise by the coalescence of subsidences is illustrated by the history of the Mediterranean" (Suess, 1909, p. 722). In an associated note, he added, "The lie of sediments which fill it is possibly synclinal, somewhat as described by Haug, Traité de Géologie, I, 1907, p. 159, fig. 36. This, however, is not the tectonic conception of the geosyncline, and the geanticline cannot be regarded as its opposite; the geanticline was originally conceived by Dana in a different sense, but to many authors both words seem to presuppose equilibrium and the germ of the isostatic theory. For this reason I regret that at first I employed the term geosyncline in this work; subsequently I avoided it" (Suess, 1909, p. 737–738). The "backthrusting" of Alpide chains around the Mediterranean is only with respect to Suess's "Asiatic structure."

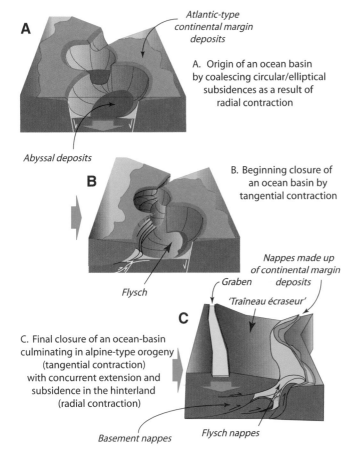

A Atlantic-type continental margin deposits

A. Origin of an ocean basin by coalescing circular/elliptical subsidences as a result of radial contraction

Abyssal deposits

B B. Beginning closure of an ocean basin by tangential contraction

Flysch

Graben *Nappes made up of continental margin deposits*

'Traîneau écraseur'

C C. Final closure of an ocean-basin culminating in alpine-type orogeny (tangential contraction) with concurrent extension and subsidence in the hinterland (radial contraction)

Basement nappes *Flysch nappes*

Figure 3. How coalescing elliptical subsidences forming ocean basins (A) may be converted into a linear and/or arcuate mountain belt (C), according to Suess's version of the theory of thermal contraction of Earth. The raceme-shaped map view shown in A is progressively converted into a narrow, linear and/or arcuate orogenic belt by progressive shortening (B and C). As far as we know, Suess was the only one who did not think that the mother basins of mountain belts had to have any map-shape similarity to their orogenic descendents. It was unfortunate that geology did not follow this important insight for nearly a century because of the blinders placed around its vision by the geosynclinal theory. The three time-lapse figures were drawn using Suess's descriptions in *Die Entstehung der Alpen* and *Das Antlitz der Erde*.

Figure 4 (below). John W. Gregory's rendition of Suess's classification of the mountains in Eurasia (from Gregory, 1915; coloring added for improved clarity). What is seen on this map as linear and/or arcuate traces had once been, in Suess's mind, ocean basins with raceme-shaped outlines probably bearing little resemblance to the map views of their orogenic offsprings. Such an interpretation was foreign to Suess's successors until well after the rise of plate tectonics. Compare this map with that shown in Figure 2, in the light of Figures 1 and 3, to have a full appreciation of how Suess imagined Tethys to have evolved into the Alpine-Himalayan mountain ranges. What Gregory here translated as "nucleus" is what Suess had called *Scheitel* and translated as "vertex" in the official English translation of *Das Antlitz der Erde*. In Suess's mind, a vertex was a focal point from which successive "waves" of mountain building emanated in the form of semiconcentric arcs.

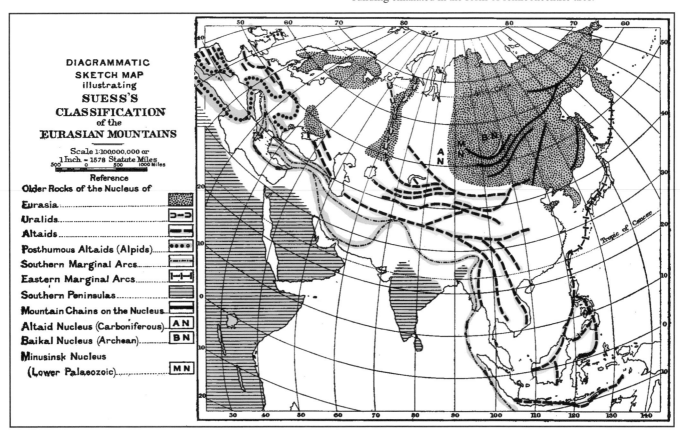

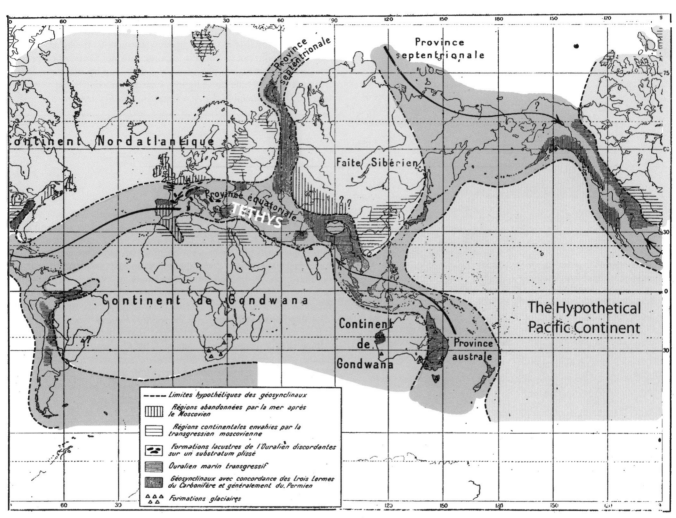

Figure 5. The paleogeography of the "Anthracolithic" (name proposed by Waagen, 1891, p. 241, for Carboniferous and Permian and adopted by Haug) world according to Haug (1908–1911, fig. 272; colored to increase clarity and rearranged by us to bring Eurasia to the middle of the map). The long, serpentine arrows indicate the migration of marine faunas. Translation of the lettering on the map: *Province septentrionale*—northern province; *Faîte Sibérien*—Siberian vertex; *Province australe*—southern province. Translation of the legend: *Limites hypothétiques des géosynclinaux*—hypothetical limits of geosynclines; *Régions abandonées par la mer aprés le Moscovien*—regions abandoned by the seas after the Moscovian; *Régions continentales envahies par la transgression moscovienne*—continental regions invaded by Moscovian seas; *Formations lacustres de l'Ouralien discordantes sur un substratum plissé*—Uralian (latest Carboniferous) lacustrine formations unconformable on a folded substratum; *Ouralien marin transgressif*—transgressive marine Uralian; *Géosynclinaux avec concordance des trois termes de Carbonifère et généralement du Permien*—geosynclines with the three parts of the Carboniferous concordant with one another and generally also the Permian; *Formation glaciaires*—glacial formations. Magenta here signifies the northern continents; khaki, Gondwana-Land; gray, the hypothetical Pacific continent; and blue, the geosynclines. Haug carried the geosyncline concept back to Europe with its new American name, but with the old Beaumontian interpretation as an intercontinental deep-sea (Haug called it "bathyal") basin, the map view of which closely resembled the mountain chain to which it allegedly gave rise. Haug had Neumayr's figure of the "young mountain ranges of the earth" (Neumayr, 1887, p. 655) before him when he drew his classical map of the Mesozoic geosynclines and continents (Haug, 1900, fig. 1), on which the geosynclines appear simply as slightly widened versions of Neumayr's mountain ranges. To obtain the map view of the late Paleozoic geosynclines, Haug widened the Mesozoic geosynclines as seen here. Haug's influence on the twentieth century tectonics was immense and, because of its geosynclinal basis, mainly negative.

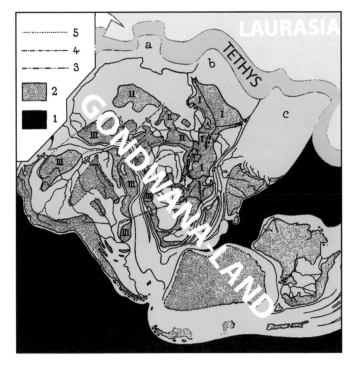

Figure 6. Argand's reconstruction of the Tethys in his classical *La Tectonique de l'Asie* (Argand, 1924, fig. 6; coloring and the names Gondwana-Land, Tethys, and Laurasia added by us to improve clarity). A shortened translation of his legend: 1. sima dominant; 2. areas in which anticlinal basement folds dominate; 3. culminations of the axes of the basement folds; 4. depressions of the axes of the basement folds; and 5. connecting lines. I, II, and III designate the first, second, and third branches of the interior virgation of Gondwana-Land. Units a, b, and c are the promontories of Gondwana-Land facing the Tethys, where a represents the African promontory, b the Arabian promontory, and c the Indian promontory. Note how Argand kept the geosynclinal, serpentine shape of the Tethys, despite the fact that he clearly had recognized that it had to widen eastward because of the constraints placed by the Pangea reconstruction. There was no reason for Argand to keep Tethys in the shape of a narrow canal mimicking the map view of its later product, the Alpine-Himalayan mountain system; in fact, he had many reasons not to do so, but his hands were tied by his commitment to the idea that mountain ranges of Alpine type arise out of geosynclines, a bias he had inherited from Haug (for a discussion of Argand's intellectual pedigree, see Şengör, 1998, p. 81–89).

of mountain ranges, which supposedly initially had the linear and/or arcuate aspect of their offspring (Suess, 1888, p. 263–264, 1909, p. 722; Suess to Ruedemann[7] quoted in Kober, 1928, p. 51). The difference between Suess's thinking and that of the geologists favoring the geosyncline idea resulted from a difference in their interpretation of the tectonic expression of the contraction theory: Suess considered oceans to result from a *radial* component of the contraction dominated by steep faults (Figs. 1 and 3A), whereas the mountain ranges built by its *tangential* component resulted in folds and shallowly dipping thrust faults (Suess, 1875, p. 146–149; Fig. 3C).

Geosyncline supporters considered both the geosyncline and the mountain range to result from horizontal shortening, i.e., from the tangential component of the contraction (Dana, 1847a; Stille, 1940, p. 10). Subsidence was supposed to weaken the bottom of the geosyncline (because it brought the geosynclinal floor closer to hotter regions in Earth), which would therefore eventually become unable to carry the contraction-driven horizontal stresses. The geosyncline would then catastrophically collapse and bear a mountain range (Dana, 1847a, 1847b). A by-product of this sort of idea was that mountain ranges are built by paroxysmal deformations expressed as "orogenic phases" (Élie de Beaumont, 1852, p. 1271; Le Conte, 1895; Haug, 1907, p. 14; Chamberlin, 1909; Kober, 1921; Stille, 1924). Because the geosyncline was essentially thought to be a giant fold formed by the same compressional stresses that eventually squashed it to generate a mountain range, it was inevitable that the axial trace of the geosyncline and the trend of the mountain range it bore had the same course.

This was foreign to Suess's thinking; he saw mountain building as a prolonged process of crustal shortening. It only stopped when a part of the shortening region fell victim anew to radial subsidence, i.e., underwent renewed basin or ocean making. Then the mountain range was said to have "solidified" (*erstarrt*; see Suess, 1875, p. 157–158, 1909, p. 720–721).[8]

When the idea of geosynclines was imported back to Europe by the great French stratigrapher Émile Haug (1861–1927) in 1900, Suess's raceme-shaped oceans were replaced by serpentine troughs winding their ways between continents (Fig. 5). Both continental drift (e.g., Argand, 1924; see Fig. 6 herein; Carey, 1958; see Figs. 7A and 7B herein) and plate tectonics inherited this, despite the fact that when Tuzo Wilson reconstructed the Pangea (Wilson, 1963) he saw, as had Boris Choubert nearly three decades earlier (Choubert, 1935) and Sam Carey in 1958, that the "Tethys" must have had a triangular shape (Fig. 8). But in a case where Wilson had no constraints, as, for example, when he was reconstructing "the ocean," the closure of which had created the Appalachians,[9] he reverted back to the geosyncline-shaped ocean interpretation (Wilson, 1966; see Fig. 9), much as had been done by Argand nearly half a century earlier (see Fig. 6 and compare the way both Argand and Wilson depicted the corresponding salients and recesses of the opposing continents).

---

[7]Rudolf Ruedemann (1864–1956), German paleontologist, who emigrated to the United States in 1892 and worked mainly on Lower Paleozoic fossils.

[8]For those who wish to learn more about Suess's ideas on global tectonics, we recommend the following papers: Şengör (1982a, 1982b, 1994, 1998, 2000, 2006).

[9]Now known to involve at least three oceans of different ages: the Iapetus, Rheic, and Theic oceans.

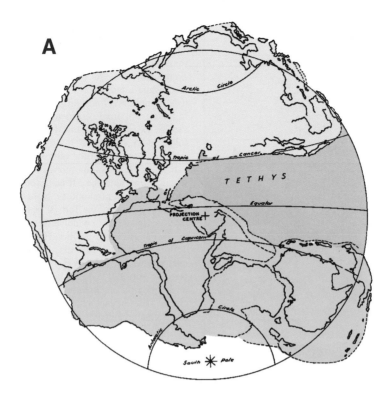

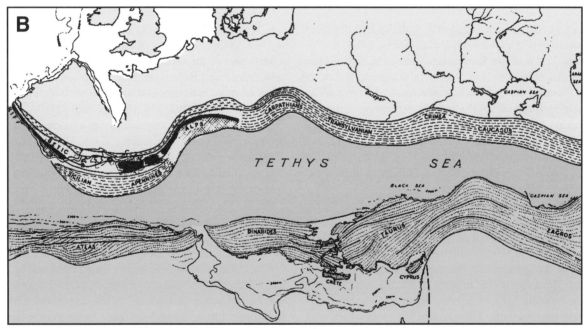

Figure 7. A: Samuel Warren Carey's 1956 reconstruction of the continents to form the late Paleozoic Pangea (from Carey, 1958, fig. 39d). Carey found the idea that a vast Tethyan gap inevitably resulted from a rigorous attempt at reconstructing the continents by closing back the Mesozoic oceans unacceptable because he "felt sure from oroclines that Indonesia and Australia belonged together" (Carey, 1958, p. 316), a shocking statement at the time for anybody who knew anything about the geology or indeed the biogeography of southeast Asia (Wallace's line!). This unfortunate choice was imposed on Carey by his commitment to geosynclines. B: Carey's preferred reconstruction of the Tethys, made by straightening out his assumed oroclines but without showing the resulting alleged "Tethyan shear" (from Carey, 1958, fig. 31b). This reconstruction ignores so many geological relationships that had long been well-established that it is not surprising that almost nobody took Carey seriously in Europe. In this map, the rotation of Spain, Corso-Sardinia, and Italy were taken from Argand (1924) without acknowledgment and the rest had so many flagrant violations of the geology of the regions depicted that it was best left out of any serious consideration. One can, however, see from the shape he gave to the Tethys that Carey was under a strong geosynclinal bias.

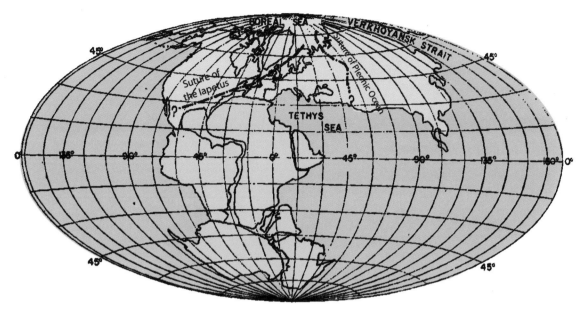

Figure 8. J. Tuzo Wilson's (1963, fig. 6) reconstruction of the continents made by closing back the Mesozoic oceans. Note that Wilson left the resulting Tethys as it appeared. We added the identification of the sutures of the Iapetus and the Pleonic oceans, which Wilson had shown, but left unidentified.

When reconstructing the evolution of a complex orogenic system involving a number of suture strands, if the shape of the former ocean that ended up bearing it is constrained by continental reconstructions around it (as in the case of the Tethys), geologists tend to invent suture-bound microcontinents that are elongated in the direction of the future orogen and freely drifting in the ocean. The partial orogens within the large, resulting orogenic system are commonly thought to be formed from whatever gets squeezed between the colliding smaller continental pieces, and these spindle-shaped pieces are assumed surrounded by ocean until the final collisional orogeny welds them to one another or to the bounding major continents (e.g., Zonenshain et al., 1990). Terranology has been one unfortunate result of this, what we call "geosynclinal psychology."[10]

Tethyan paleogeographic reconstructions have suffered much from this sort of cross-sectional, concertina tectonics, as Tethyan oceans have often been depicted as trough like as possible, and whenever a continental piece has been discovered in the Tethyan realm, it has been drawn as if it had been a completely independent island to the last, so as to yield as many serpentine troughs between various continental pieces as possible. One of us has been as much responsible for this sort of "cylindristic" thinking as anyone (e.g., Şengör et al., 1984, 1988; Şengör and Hsü, 1984; Nie et al., 1990; Flügel and Hubmann, 1993; Ricou, 1994, 1996; Şengör and Natal'in, 1996; Ziegler et al., 1997). More recently, even more extreme cylindrist interpretations have appeared, such as those by Stampfli et al. (1991, 2001).

However, four groups of development soon showed that all such reconstructions must be unlikely because:

(1) Paleomagnetic and sedimentary facies data made clear that many east-west features in the Tethyan realm had not always had that orientation. In fact, many had gone through more than 90° rotations around vertical axes during the Tethyan evolution (for a review see Şengör, 1990c,[11] 1990d; Van der Voo, 1993; Burtman, 1994).

(2) There was much strike-slip faulting within the Tethyan realm and in its surroundings making it impossible to reconstruct a simple open-and-shut case for most segments of the Tethyan oceans (e.g., for a listing, see Şengör et al., 1988; also see Şengör,

_____

[10]Our readers will notice our complete avoidance of both the "terrane" terminology and the associated recommended methodology. Şengör (1990b) and Şengör and Dewey (1990) discuss the reasons behind our choice.

[11]Şengör (1990c) had relied on data available prior to 1990 to reconstruct the tripartite subdivision of the so-called Central Iranian Microcontinent (Takin, 1972), consisting of the Yazd, Tabas, and Lut blocks and their counterclockwise rotation during the Mesozoic. Since then the results of the immensely detailed and superb facies studies of the Devonian and the Lower Carboniferous of Iran by Wendt et al. (1997, 2002, 2005) have been published, in which the authors do not see any necessity for the counterclockwise rotation of the central Iranian blocks. Their data, however, are not incompatible with the rotation, which is independently required by the available paleomagnetic data (accepted by Van der Voo, 1993, p. 193–194) and by the distribution of the late Paleozoic basement provinces (see Şengör, 1990c. Wendt et al. [2005, p. 58] accept the late Paleozoic to Triassic metamorphism of the rocks in the Anarek area.). Moreover, the bounding faults of the Yazd, Tabas, and Lut blocks, namely those of Posht-e Badam, Kuh-e Kalshaneh, Kuh-e Banan, and Nayband, remained active for too long throughout the Phanerozoic in different roles to have been located in a quiet continental interior as the interpretation of Wendt et al. would require. In fact, Wendt et al. (1997, especially fig. 21) themselves have added to the data bank of synsedimentary faults three superb, steeply-dipping faults along the Kuh-e Kalshaneh structure during the Devonian. For such reasons, we here adhere tentatively to the interpretation of Şengör (1990c) in terms of the late Paleozoic to late Cenozoic tectonics of Iran.

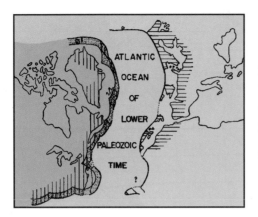

Figure 9. J. Tuzo Wilson's reconstruction of "the ocean," the closure of which had supposedly created the Caledonides and the Appalachians (from Wilson, 1966, fig. 3). Note that with its almost perfectly parallel margins, it looks geosynclinal. Geosynclinal bias of a scientist who had already rejected the idea of geosynclines?!

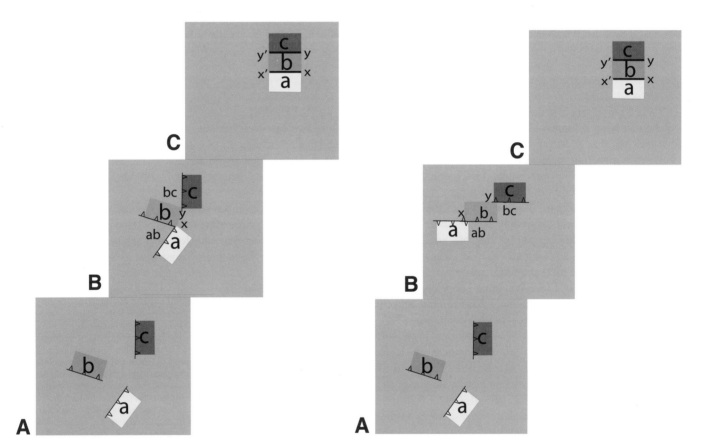

Figure 10. Sketch maps showing the difference between "contact" (B) and "terminal collision" or "suturing" (C). In A, the three continents a, b, and c are not in any sort of continental contact with one another. In B, they come into contact with one another at the corners x and y. In C, they have terminally collided (sutured) by eliminating the oceanic surfaces ab and bc between a and b, and b and c, respectively, and formed the sutures y'y and x'x. Transition from the contact stage to terminal collision stage in this case involved significant rotations around vertical axes. There are many examples of such events within the Tethysides.

Figure 11. Sketch maps showing the difference between "contact" (B) and "terminal collision" or "suturing" (C), whereby the sutures y'y and x'x are formed mainly by strike-slip emplacement of one continent with respect to the other instead of rotation of the colliding continents with respect to one another around a pole contained at one end of the terminal suture. These kinds of sutures have been called "transform sutures" by Dewey et al. (1986). The Xiangganzhe suture in southern China is a possible example of such a suture. Transform sutures may also involve substantial rotations of their bounding continents around vertical axes, similar to a shearing card deck.

1990c, 1990d, 1993; Görür and Şengör, 1992; Natal'in and Şengör, 2005). Despite that, most reconstructions have continued to show the Tethyan realm consisting of one vast triangular, eastward opening gulf of the Pangea populated by numerous, small, independent continental fragments.

(3) The evolution of Tethyside arcs display changes in past behavior suggesting complex plate boundary evolution within the Tethyan oceans incompatible with simple open-and-shut scenarios (e.g., Şengör et al., 1991, 1993a; McQuarrie et al., 2003; Natal'in and Şengör, 2005).

(4) Many intra-Tethyan continental blocks appear to have had land connections between one another or proximities indistinguishable from actual land bridges clearly before the sutures that now weld them closed. This suggests that contact between them had occurred in a geometry not now reflected by their union across their mutual suture(s). For example, continental pieces making up China and Indochina east of the Nan-Uttaradit–Sra-Kaeo–Bentong-Raub suture (Şengör, 1986; Şengör et al., 1988) share a Cathaysian flora already in the late Carboniferous and a benthic foraminiferal fauna characterized by the genera *Neoschwagerina*, *Lepidolina,* and *Paleofusulina* in the mid to late Permian (see Şengör et al., 1988). In addition, it seems that in the late Permian, the clumsy little mammal-like reptile *Dicynodon* somehow managed to reach Indochina from Gondwana-Land and from there it went to eastern Laurasia (Battail, 1997, 2000). It is therefore most unlikely that the areas now making up this portion of eastern Asia could have been separated by wide oceans in the latest Paleozoic. *Yet the major sutures separating them are demonstrably Triassic structures!* This is only possible if these blocks came into contact in geometries very different from their final welding geometries. One must thus distinguish a "contact time" from a "terminal collision time" when considering paleogeographic reconstructions of continental entities (Figs. 10 and 11).

By late Permian time, however, many of the intra-Tethyan continental fragments had already contacted one another and the major bounding continents of Laurasia and Gondwana-Land; thus it was not possible for them to be free-floating continental platforms and/or magmatic arcs within a late Paleozoic Tethys as commonly depicted (e.g., Şengör et al., 1988; Görür and Şengör, 1992; Ziegler et al., 1997). *Yet, along much of its east-west extent, the late Paleozoic Tethys did remain wide open until the late Triassic and in many places even until the medial Jurassic, because terminal collisions along the Cimmeride sutures did not occur until then.* It is clear that the map shape of the Paleo-Tethys had not the remotest resemblance to the mountain ranges that eventually grew out of its subductive removal and final collisional destruction. How to reconcile a wide-open Tethyan ocean with a string of continents in contact with each other and with the main Tethys-bounding continents at its eastern, widest end constitutes a major paleogeographic problem that needs to be solved.

To our knowledge, our assiduous friend Chris Scotese (2001) was the first to arrange the intra-Tethyan continents in such a way as to allow *early contacts* and *much later terminal collisions*. Others followed him shortly thereafter (e.g., Golonka, 2002; Stampfli and Borel, 2002). However, Scotese has obviously remained under the influence of the earlier free-floater depictions that prevented him to take the final logical step and to proclaim the Paleo-Tethys an essentially closed sea, a continent-locked ocean, an independent basin from the Panthalassa, much like the Indian Ocean's independence from the Atlantic on Claudius Ptolemy's maps from the second century AD (Stevens, 1908; Fischer, 1932a, 1932b; Figs. 12A and 12B). Indeed when one looks at the geology of the sutures connecting the various continental blocks now embedded in the Tethyside collage, one is struck by the fact that all the pieces in China and Mongolia had already established their initial contacts with one another by the late Permian, and only a very narrow strait in northern Thailand (Golonka, 2002; Stampfli and Borel, 2002) may have separated them from the easternmost representative of the Cimmerian continent, the Sibumasu of Ian Metcalfe (1996), including the Shan Plateau of Burma and what the late, regretted Luc-Emmanuel Ricou called the Axial Thai Gneiss Belt (Ricou, 1994, 1996).

It was in the course of establishing the paleogeography of the late Permian Paleo-Tethys that we became aware of the early onset of oxygen-poor conditions within it and that its unique paleogeographic conditions must have been the cause of that onset.

Before we proceed with the paleogeography of the Tethyan realm in late Paleozoic time, we need to state the terms commonly used in Tethyan geology for those not familiar with them.

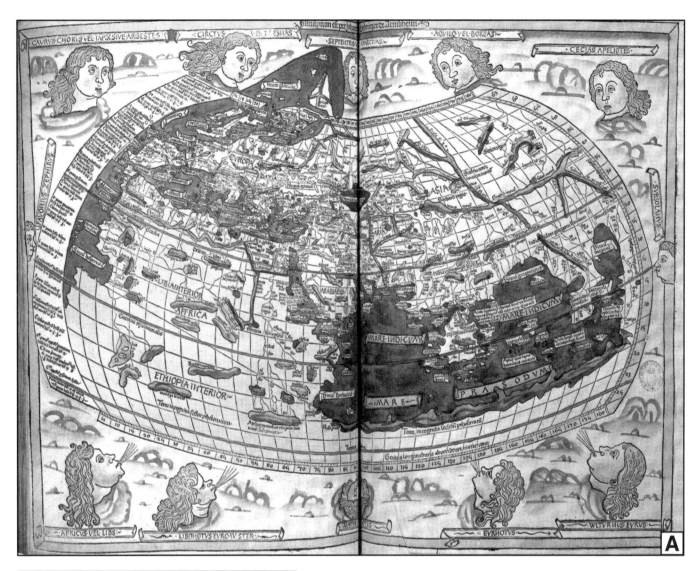

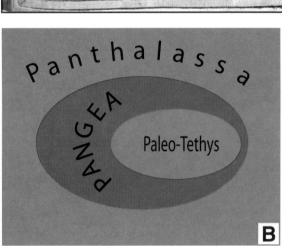

Figure 12. A: Ptolemy's *mappa mundi* from the 1482 Ulm edition of his Γεωγραφιχη Υφηγησις (*Geografike Ufegesis* or Guide or Manual of Geography) dating from the second century AD (Ptolemy, 1482). Note the completely enclosed geography of the Indian Ocean (*Mare Indicum* and *Mare Prasodum*, i.e., the "Green Sea"). The idea of a completely closed Indian Ocean may date back to Hipparchus (*floruit ca.* 140 BC) on the basis of Seleucus's (358–281 BC) report of the independence of the tides there from those of the Mediterranean (see Dicks, 1960, p. 114–115). B: A conceptual depiction of the Ptolemaic condition on the idealized example of the late Permian Pangea, the Paleo-Tethys, and the Panthalassa. Note the total isolation of the Paleo-Tethys within a continent similar to the isolation of the Indian Ocean in Ptolemy's world map. This is what we call the complete Ptolemaic condition. A partial Ptolemaic condition is achieved when only a few, narrow, and bathyal to neritic passages, sufficient to inhibit the outflow of heavy waters, connect the enclosed ocean with the outer ocean, which was actually the geometry in the case of the Paleo-Tethys for most of its existence.

CHAPTER III

The Tethyan terminology

The following definitions refer to the elements illustrated in Figures 13 and 14.

1) The *Tethyan realm* refers to the paleogeographic area covered by all the Tethyan oceans and their continental margins, including any associated magmatic arcs and any continents located within and between the Paleo- and the Neo-Tethys. It may or may not coincide with various Tethys-based paleozoogeographical provinces and/or realms and should not be confused with them.

2) *Paleo-Tethys* is the remnant ocean that originated after the final assembly of Laurasia (Laurussia and Siberia) at the expense of the Khanty-Mansi (Şengör et al., 1993a) and Pleonic (McKerrow and Ziegler, 1972) oceans during the late Carboniferous to the late Permian interval and the final collision of Laurasia with Gondwana-Land during the medial Carboniferous by the elimination of the Rheic and the Theic (in the French literature also known as *Océan Centralien*: e.g., Michel, 2008) oceans (McKerrow and Ziegler, 1972). In the place of these oceans the Altaid (Khanty-Mansi), Uralide (Pleonic: McKerrow and Ziegler, 1972), Hercynide (Rheic and Theic), and the Great-Appalachide[12] (final formation included elimination of the Theic Ocean) mountain belts formed. The Scythides (Natal'in and Şengör, 2005) formed north of the Paleo-Tethys and connected the Uralides and the Hercynides. Unlike the latter two orogenic belts, their orogenic evolution merged with that of the Cimmerides, and orogeny ceased along them only after the final elimination of the Paleo-Tethys in the Triassic to early Jurassic interval (Natal'in and Şengör, 2005).

3) The *Cimmerian continent* is the continental strip that rifted off the northern margin of Gondwana-Land to close the Paleo-Tethys before it and to open the Neo-Tethys behind it (Şengör, 1979; see also Figs. 15B and 15C). This happened during the early Carboniferous to

medial Jurassic interval, when the Cimmerian continent rotated counterclockwise around a pole somewhere in present-day Bulgaria. The only discontinuity we can find in this strip-continent to divide it into two pieces is the rift and/or ocean system extending from the Pamirs, via the Wašer and Sistan oceans in Afghanistan and eastern Iran, respectively, to Oman (see especially Gaetani, 1997a and 1997b).

4) The *Cathaysian bridge*, a term we introduce here, designates a broad continental isthmus consisting of Annamia, the Huanan block, the Yangtze block, and the North China block, together with its Manchuride appendage (Şengör and Natal'in, 1996). It constituted a veritable "isthmian link," very much in the sense of Bailey Willis (1932), between eastern Gondwana-Land and eastern Laurasia during the medial to late Permian.

5) *Neo-Tethys* is the ocean that opened between Gondwana-Land and the Cimmerian continent between the early Carboniferous in northwestern Australia and western Thailand and early Jurassic in the Southern Alps and northern Calcareous Alps. In Figure 13, we have depicted the Alpine, Apennine, Tellian Atlas, Rif, and Betic sutures in red implying that they represent oceans originally part of the Atlantic Ocean. This has long been recognized ever since the pioneering studies by Hsü (1971), Smith (1971), Bernoulli (1972), Dewey et al. (1973), and Bernoulli and Jenkyns (1974), and that is why Şengör (1984, p. 65) even made the suggestion to separate all the orogenic belts classically regarded as "Tethyside" west of the Carpathians and the Dinarides as products of the Atlantic Ocean and to call them "Atlantides" rather than "Tethysides."

We here define a sixth, but general, term, independent of the Tethyan realm, namely the *Ptolemaic condition*. It refers to a state of geography on Earth in which two or more oceans coexist with no contiguous, normal oceanic floor between them (Fig. 12B). The existence of the Mediterranean and its satellite oceanic basins such as the Black Sea impart onto our present-day world a Ptolemaic condition. A *complete Ptolemaic condition* is one in which two or more oceans are so completely isolated from one another that not even any aqueous communication between them is possible. The South Caspian ocean provides an example of the complete Ptolemaic condition on our planet today. The South Atlantic at times during the Aptian (Burke and Şengör, 1988) and the Mediterranean during the evaporative drawdown phase halfway through the Messinian salinity crisis (Ryan, 2009) displayed complete Ptolemaic conditions. The Para-Tethys (Laskarev, 1924) also was a Ptolemaic ocean alternating throughout its

---

[12] The concept of *Großappalachiden* (Great-Appalachides) was introduced by Hans Stille in 1940 (p. 33) to embrace both the Appalachians and the Ouachitas plus the other smaller inlier, in the Glass Mountains of the Marathon area in Trans-Pecos Texas, of the great late Paleozoic collisional orogen that flanks North America to the east and southeast. It is now recognized that this orogenic system continues southward via the Huastecan deformations in Mexico (de Cserna, 1960). The Huastecan belt corresponds to the Coahuiltecano and Tarahumara "terranes" of Sedlock et al. (1993) and clearly continue in Coahuila, Chihuahua, Sinaloa, and Sonora (Poole et al., 2005). The Great-Appalachides also continue eastward into the Mauritenides of western Africa (Sougy, 1962; Villeneuve, 2008). The Great Appalachides end in northern Mexico against a large transform fault boundary that transferred the shortening out into the Panthalassan margin and may thus have triggered the Sonoma Orogeny in the North American Cordillera in a manner not dissimilar to the suggestion by Stevens et al. (2005).

history since the Priabonian–Beloglinian (late Eocene) between complete and incomplete Ptolemaic states (Popov et al., 2004). *All these isolated oceans had one or more episodes of anoxia or at least hypoxia during their histories* (e.g., for the South Atlantic, see the references cited in Weissert, 1981; for the Mediterranean, see Ryan and Cita, 1977; for the Para-Tethys, see Schmid et al., 2001). The existence of the pre-Caspian depression as a completely landlocked ocean since the Permian, in which vast salt deposits of that age accumulated, has imposed a Ptolemaic

condition onto Earth ever since, because it still remains unclosed, as Kevin Burke showed more than 30 yr ago (Burke, 1977; see also Şengör and Natal'in, 1996). A similar condition, much disputed later, was suggested by Hallam (1969) to account for the origin of the peculiarities of the Jurassic boreal fauna.

Two or more major oceans forming a Ptolemaic world has not been an uncommon condition in Earth history, and we think that the late Permian did have a serious Ptolemaic condition as we document below.

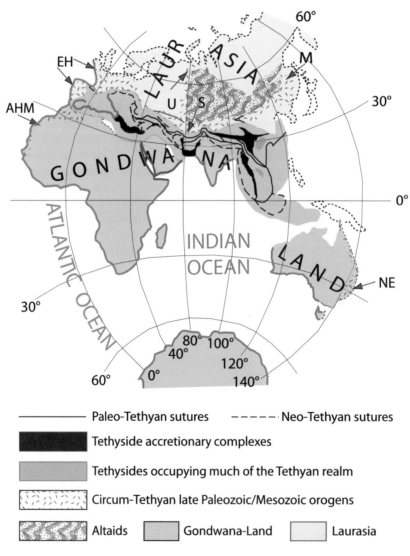

Figure 13. The late Paleozoic to present large-scale structure of the Old World showing the position of the Tethyan realm within it. The rifted continental margins of the two different oceans disrupting the Old World are shown in the colors corresponding with the color in which their names are written (except the Red Sea and the Gulf of Aden, whose names could not be written because of the exigencies of the available space). The colors and patterns in this figure correspond to those in Figures 14. AHM—African Hercynides and Mauritenides; EH—European Hercynides; M—Manchurides; NE—New England foldbelt. For the meaning of the colors of lines depicting continental margins, see the legend to Figure 14.

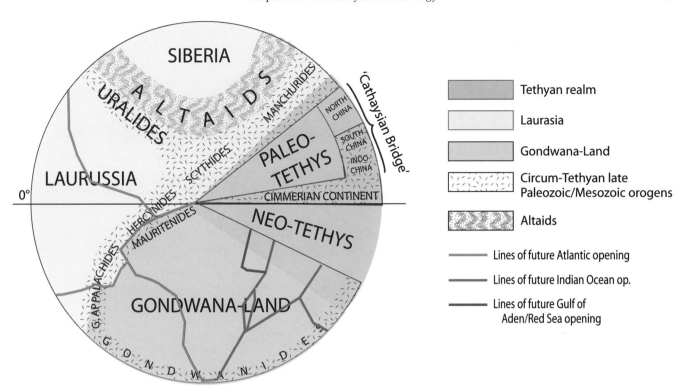

Figure 14. A schematic depiction of the main paleotectonic elements of the world in late Permian time illustrating their terminology used in this paper. The colors and patterns in this figure correspond to those in Figure 13.

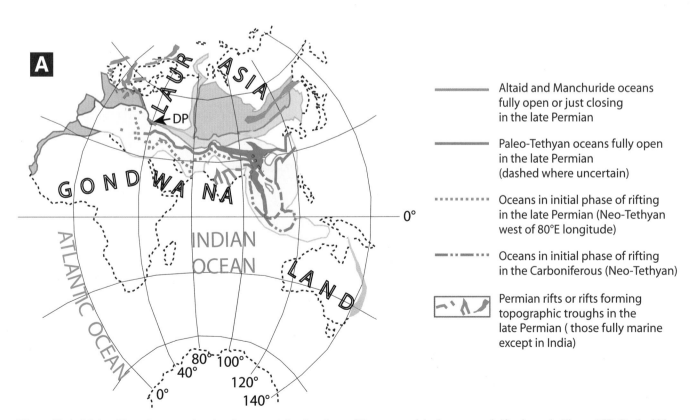

Figure 15. A: Major rifts and sutures showing the present-day locations of the now vanished oceans and rifts shown in Figure 15B. Dashed blue suture is pure conjecture. DP—Porte of Dobrudja. (*Continued on following pages.*)

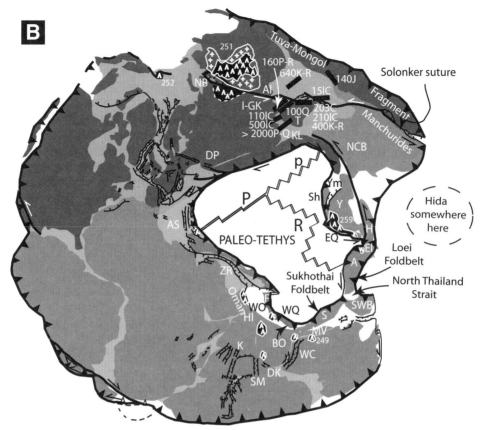

Figure 15 (*continued*). B: Late Permian tectonics of Pangea and the Paleo-Tethys. Key to lettering: A—Annamia; Al—Altay Mountains; AS—Apulian shelf (African promontory of Argand, 1924); BO—Banggong Co-Nu Jiang Ocean; DK—Damodar-Koel Valley rift; DP—Porte of Dobrudja; El—Emei Shan basalts on Annamia; EQ—eastern Qangtang; F—Farah block; H—Huanan block; Hl—Helmand block; I-GK—Irtysh-Gornostaev keirogen; K—Kurduvadi rift; KL—Kuen-Lun; MV—Mount Victoria Land block; NB—Nadym Basin; NCB—North China block; S—Sibumasu; Sh—Shaluli Shan ensimatic island arc; SM—Son-Mahanadi rift; SWB—south-west Borneo; T—Tarim block; WC—Western Coastlands taphrogen; WQ—western Qangtang; Y—Yangtze block; Ym—Yajiang marginal basin; ZR—future Zagros Mountains. Lone figures are isotopic ages (in Ma) of extensional and/or plume-related igneous rocks, mostly traps. The figures with letters after them represent observed amounts (in km) and times of crustal shortening across the bars associated with them: >2000P-Q—Permian to Quaternary: Rzhevsky and Khramov (1985); 160P-R—Permian to Recent: Burtman et al. (1996); 140J—Jurassic: Dumitru and Hendrix (2001); 640K-R—Cretaceous to Recent: Y. Chen et al. (1992); 400K-R—Cretaceous to Recent: Y. Chen et al. (1992); 203C—Cenozoic: Avouac et al. (1993); 500lC—late Cenozoic: Métivier and Gaudemer (1997); 110lC—late Cenozoic: Cobbold et al. (1996); 15lC—late Cenozoic: Cunningham et al. (2003); 210lC—late Cenozoic: Yin et al. (1998); 100Q—late Quaternary: Thompson et al. (2002). Sources for rifts: main database and Şengör and Natal'in (2001); supplemented by: for the Dinarides: Herak (1999); for the Hellenides: Bernoulli and Laubscher (1972); for the Iberian Peninsula: López-Gómez et al. (2002); for details in northern Spain only: see Martinez Garcia (1990); for the North Atlantic rifts: Ziegler (1988, 1990); for Tunisia: Solignac and Berkaloff (1934),[13] Busson (1969), and Ben Ferjani et al. (1990, p. 33–35); for the Zagros–southeastern Turkish area: Şengör (1990c); for Turkey in general: Şengör and Yılmaz (1981). Sources for subduction zones: for Australia: Veevers (2000, 2001); for the Tethyan realm: Şengör et al. (1988) and Şengör and Natal'in (1996); for the North American Cordillera: Burchfiel et al. (1992a), Miller et al. (1992), Plafker and Berg (1994), and Dickinson (2004); for Mexico: Keppie (2004) and Nance et al. (2006); for the Arctic: Natal'in et al. (1999) and Boris A. Natal'in (2007, personal commun.). Sources for the traps and other rift volcanics: Canning Basin, north-west Australia: Reeckmann and Mebberson (1984); Carnarvon Basin, north-west Australia: Bradshaw et al. (1988); Panjal: Honegger et al. (1982), Searle (1983), and Vannay (1993); Provence: Cassinis et al. (1995); Sikkim: Sinha-Roy and Furnes (1978, 1980) and Rao and Rai (2007).

---

[13]A superbly detailed study with many fossil reports and worldwide comparisons.

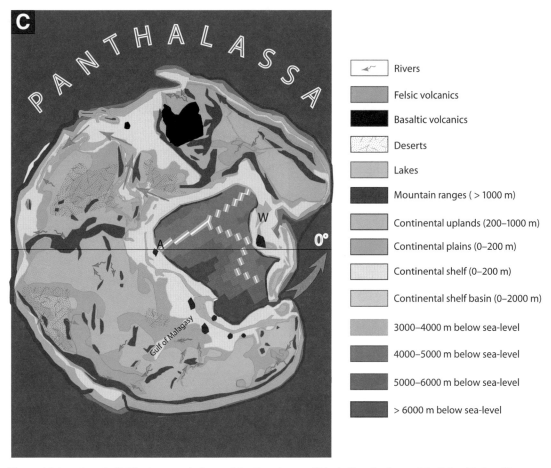

Figure 15 (*continued*). C: The geomorphology of the Permian world including the floor of the Paleo-Tethys. No morphological elements are shown within the Panthalassa except the inferred circum-Panthalassan trench, which, in fact, may not have existed as a topographic feature everywhere. A is where the Aksu Flysch in northern Turkey is seen; W is Wolong in western Sichuan (ancient Xikang). The sources are as follows: shorelines: main database is Ziegler et al. (1997); supplemented by: for northwest Australia: Bradshaw et al. (1988); for northeast Africa: Schandelmeier et al. (1997); for the North Atlantic-Arctic–European region: Ziegler (1988, 1989, 1990); for east Africa-India: Veevers and Tewari (1995). Sources for land topography: main database is Ziegler et al. (1997); supplemented by: for Australia and Antarctica: Veevers (2001). Age of Whitehill "sea" in South Africa and South America is from MacRae (1999, p. 172). Sources for rivers: for Australia and South Africa: Veevers (2001); for India: Casshyap (1982) and Veevers and Tewari (1995); for northeast Africa: Schandelmeier et al. (1997); for North America: Johnson et al. (1988); for South Africa and Australia: Veevers (2000, 2001); for northeast Siberia: Khudoley and Guriev (1994).

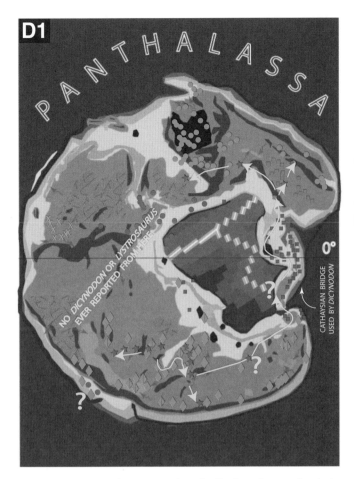

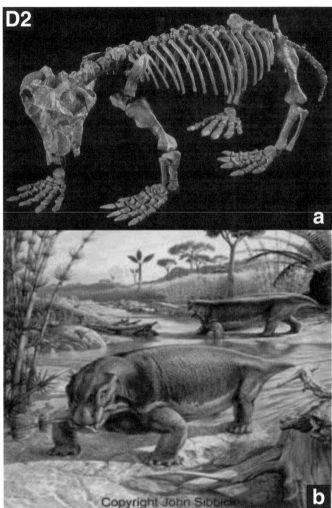

Figure 15 (*continued*). D1: Permian distribution of some floral elements showing provinciality. Blue indicates cold climate plants; circles represent the Angara flora, whereas the diamonds represent the *Glossopteris* flora. Squares are mild to tropical climate floras, called Cathaysian. The orange squares are the north Cathaysian elements, whereas the red ones represent south Cathaysian elements. The magenta circles are the elements of the Euramerian flora and the orange X's represent the Southwestern United States flora. The map was compiled from Meyen (1970), Chaloner and Meyen (1973), Lacey (1975), Chaloner and Creber (1988), and Şengör et al. (1988). This map shows that plant distribution in the Permian was almost entirely climate controlled and not so much isolation controlled, as Ziegler (1990) correctly emphasized almost two decades ago, because there were almost no isolated pieces of land. The presence of mixed floras corroborates this inference. The two mixed Gondwana-Land–Euramerian province localities in Patagonia reported by Lacey (1975) come from regions not parts of South America at the time. As we do not know where those regions were (except that they were most likely parts of Gondwana-Land), we plotted them on the most southerly position possible in South America, but with a question mark. The green stars are places where *Lystrosaurus* sp. fossils were found. As the earlier *Dicynodon* representatives (Buffetaut, 1989; Battail, 1997), the *Lystrosauri* must also have used the Cathaysian bridge to reach Laurasia (yellow arrows) because their distributions are so similar. Note that southern Cathaysian floral elements show up in the Middle East at similar latitudes as the Chinese and southeast Asian elements. This corroborates the presence of the Cathaysian bridge and the climate control of the distribution of the floral elements.

Figure 15 (*continued*). D2: *Lystrosaurus* sp. A shows a skeleton copied from the Internet site http://www.math.montana.edu/~nmp/materials/ess/geosphere/inter/activities/exploration/lystr.jpg on 8 January 2008. B exhibits a full reconstruction by John Sibbick emphasizing the cylindrical body, including the huge chest, here reproduced with his kind permission.

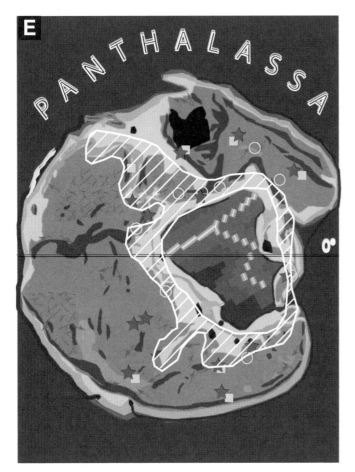

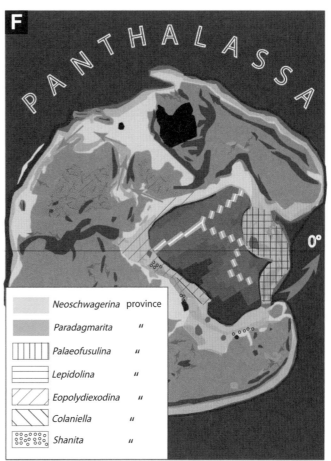

Figure 15 (*continued*). E: Tetrapod distribution during the late Permian (Platbergian) and early Triassic (Lootsbergian) times. Permian localities (red stars) are from Lucas (2006) and the Triassic localities (yellow squares) are from Lucas (1998), where each locality represents a general area, such as "western Europe" or "Karoo basin." Empty, yellow-rimmed circles represent early Triassic terrestrial sedimentary rocks in facies appropriate for finding tetrapod fossils and where none so far have been reported. The white ruled area represents the regions where anomalous abundances of fungal remains of latest Permian age are reported (see Fig. 24).

Figure 15 (*continued*). F: Distribution of late Permian benthic foraminifera from the Tethyan realm. Note the similarity of the *Neoschwagerina* province to that of the south Cathaysian floral kingdom. The benthic foraminifera also show that by late Permian time the Cathaysian bridge was complete, at least intermittently. The foraminiferal distribution is from Şengör et al. (1988).

CHAPTER IV

The tectonics of Pangea and the Paleo-Tethys during the late Permian

It cannot be the task of this section to present a complete discussion of the tectonics of the late Permian Pangea and the Paleo-Tethys it enclosed. The subject is too vast to be compressed between two book covers. Rather, we aim to establish that in the late Permian the Paleo-Tethys was indeed cut off from the Panthalassa by the Cathaysian bridge and to discover its internal bathymetry. Since the first-order oceanic bathymetry is a function of the age of the ocean floor, we try to see whether we can establish the geometry of the ridges in the Paleo-Tethys.

Figure 15A shows in a general way the sutures that formed from the late Permian oceans and the late Permian rifts, some of which later became oceans on the present face of Earth. Figure 15B is a tectonic map of the late Permian Pangea illustrating what those elements looked like then and some other tectonic elements relevant to the following discussion. Figures 15C through 15F show various physical geographical and biogeographical elements underpinning our reconstruction of Pangea.

The northern margin of the Paleo-Tethys (Figs. 15B and 15C) was shaped largely as a consequence of the Altaid, Scythide, and Manchuride deformations (Fig. 14; Şengör and Natal'in, 1996, 2004; Allen et al., 1995; Natal'in and Şengör, 2005). By late Permian time, left-lateral displacement along the Irtysh-Gornostaev keirogen (I-GK in Fig. 15B) had brought the two moieties of northern Asia it divided almost to their present-day positions relative to one another, as shown in Figure 15B. The northern end of the Irtysh-Gornostaev keirogen is located in the deep Nadym basin (NB in Fig. 15B), which is the largest and the deepest individualized depocenter of the west Siberian depression. Here "late Paleozoic" deposits reach a thickness of 5 km and are located under a 5–9 km thick sediment cover belonging to the larger west Siberian basin complex (Şengör and Natal'in, 1996). The depth to basement in the Nadym basin, the mafic nature of its basement, and the absence of unconformities in its fill led Aplonov (1989) to consider it a small ocean. We think that he may not be very far from the truth in that the Nadym probably represents a large amount of extension, possibly absorbing a good portion of the 2000 km Permian offset along the Irtysh-Gornostaev system (Şengör et al., 1993a; Allen et al., 1995; Şengör and Natal'in, 1996). It is remarkable that the largest plateau basalt eruption area in the Phanerozoic, the Tunguska trap province in Siberia (the large volcanic province between NB and 251 in Fig. 15B; 251 is the putative main eruption time, in Ma before present, in the province), lies athwart the Nadym basin and the Irtysh-Gornostaev keirogen farther south. The total volume of the trap basalts and associated igneous rocks has been estimated anywhere between 2 million km^3 (Milanovskiy, 1976) and 4 million km^3 (Masaitis, 1983). The time of eruption is often expressed to have been less than 1 Ma, but this is most unlikely in the face of the paleontological evidence from the intratrappean sediments suggesting a duration of some 5 Ma[14] (Hallam and Wignall, 1997, p. 136). This is corroborated by the most recently published compendium of isotopic ages we can find at http://www.le.ac.uk/gl/ads/SiberianTraps/Dating.html (seen on 15 December 2007) that span an age interval of some 4 Ma.

If one considers a 3300-km-long fast-spreading ridge (roughly the north-south extent of the Tunguska trap province) with a spreading rate of 16 cm/a (about the rate of an ultrafast spreading ridge today[15]), it would generate an oceanic crust of some 2,640,000 km^3 in one Ma (assuming a crustal thickness of 5 km). If the duration is extended to 5 Ma, the volume would grow to 13,200,000 km^3. The amount of offset along the Irtysh-Gornostaev keirogen was 2000 km, and this was accomplished sometime during the Permian. Let us say that the extension took the entire time represented by the Permian (~50 Ma and the rate of extension we get is thus 20 cm/a, equal to the fastest spreading known on Earth: see footnote 15 above) and that the eruptions occupied only the last 5 Ma. This could give us an offset of some 400 km if we assume the rate of motion along the Irtysh-Gornostaev keirogen to have been uniform. If this were all spreading along a ridge 3300 km long, the volume of oceanic crust generated would have been some 13,200,000 km^3. If we reduce the spreading rate[16] or the time interval of eruptions by four times, we still get some 3,300,000 km^3; if we also halve the ridge length

---

[14]Could be as short as 4 Ma according to the most recent calibration of the early Triassic: see Brayard et al. (2007).

[15]The fastest spreading we know to have occurred on our planet was some 20–22 cm/a some 20–11 Ma ago on the superfast East Pacific rise, where the Cocos plate is separating from the Pacific plate (Teagle and Wilson, 2007). We are grateful to Prof. Joann Stock of Caltech for this piece of information.

[16]The 2000 km offset along the Irtysh-Gornostaev keirogen is derived from our Altaid reconstructions (Şengör and Natal'in, 1996), which could easily tolerate a margin of error of some 500 km, but not much more. So it is difficult to cut the rate of motion along the Irtysh-Gornostaev keirogen by half. We seem to be stuck with a rate of at least 15 cm/a, which is similar to the ultrafast spreading rates of the present. This is faster than any continental transform fault boundary today. However, regarding the rates of evolution of the basins along the Irtysh-Gornostaev system as a whole, this is not surprising. In fact, such an ultrafast spreading rate within the Tunguska extensional province may be the explanation of the phenomenal amounts of basalt produced in the province. The "Siberian traps" are a unique phenomenon during the Phanerozoic. The associated rate of extension in a continental milieu is probably also unique and in part may be the explanation of the uniqueness of the igneous province. However, we emphasize that the rate of extension one gets is similar to the known rate of spreading in the superfast East Pacific rise in the equatorial Pacific.

we get 1,150,000 km³, which is near the lower estimate for the volume of the Tunguska trap volume. Now, what if we distribute the eruptions to many conduits instead of confining it to a singularity along a ridge and honor the paleontological dating? In other words, instead of making sheeted dykes along a single continuous ridge, we make a dyke swarm associated with extension related to regional, broad strike-slip represented by the Nadym and other associated much smaller basins along the Irtysh-Gornostaev keirogen in a time span of some 4 Ma? We can reach the higher estimates for the amount of basalt only by lithospheric stretching without the help of a mantle plume. We reserve the discussion of the Tunguska traps for a different paper, but here simply note that their production was probably not confined to a time interval less than a million years and therefore may not necessarily have required a mantle plume as has been frequently assumed.

Farther south and east we come to a region where the Altaid orogenic system meets the north China block (NCB in Fig. 15B) and its (present) northern accretionary complex, the Manchurides (Şengör and Natal'in, 1996; Natal'in, 2007). Very large amounts of intracontinental shortening have gone on in regions surrounding the stiff Tarim block, and it is still going on. We have assumed a conservative 2000 km maximum shortening within this zone (shown in lighter gray shade in Fig. 15B surrounding the Tarim block, T, between the late Permian and Recent). This is based on the simple inference that since India collided with Asia, some 2000–2500 km intracontinental shortening occurred between cratonic India and cratonic Siberia as indicated by paleomagnetic data (see fig. 9 in Chen, 1992). Some 1000 km of it has been absorbed in the Himalaya and the rest distributed in Asia north and east of the Himalaya (Chen, 1992, fig. 9). We get a rough idea of where and how this is distributed by looking at the present day strain field in Asia (e.g., Holt et al., 2000; Flesch et al., 2001). The present strain field gives us essentially an instantaneous picture. In order to extend that picture into a finite past, in Figure 15B we have given examples of quantitative estimates of shortening and the time intervals in which that intracontinental shortening occurred wherever we could find data north of the Tibetan Plateau. From these, one gets some 150 km of shortening in the Tien Shan in ~10 Ma since the India-Asia collision, which is too high a rate. Since that collision commenced ~55 Ma ago, we would thus have obtained ~750 km shortening by assuming uniform rates since the collision. Y. Chen et al. (1992) estimate some 400 km shortening since the Cretaceous with an uncertainty larger than the estimate. Still, their estimate, without the uncertainty, makes sense in view of the geology and average plate motion rates. This is not unrealistic, also because deformation spread northwards in ever increasing rates from Tibet. We further assume that in other places such as the Kuen-Lun and the Altay similar shortening occurred. This is justified in view of the estimate of Y. Chen et al. (1992) of some 640 ± 530 km shortening north of Junggar. We thus note that ~750 km overall shortening in areas surrounding the Tarim basin since the India–Eurasia collision is not a fantastic figure.

The character of intracontinental deformation in central Asia after the Cimmeride collisions was very similar to the situation after the India-Asia collision (Şengör, 1984, 1987). Zhang et al. (2007) recently made the astonishing discovery that ultrahigh pressure metamorphism in the western Tien Shan may have lasted until the beginning of the Carnian, and they even claim a Triassic age for the terminal collision here. We doubt whether the collision could have been that late because the stratigraphy indicates termination of oceanic conditions in the Permian. The subduction discovered by Zhang et al. (2007) may have been of the intracontinental type, similar to the one active beneath the Pamirs today (Roecker, 1982), some 55 Ma after the terminal collision. But, however one interprets the data, large amounts of shortening north of the Tarim block during the Triassic seem unavoidable. If it also produced about the same amount of shortening as did the later India-Asia collision, one can see that a rough estimate of 2000 km for the Permian to Recent shortening in central Asia is a very plausible amount. Indeed, qualitative statements of intense intracontinental shortening between the Altay Mountains (Al in Fig. 15B) and the Kuen-Lun ranges (KL in Fig. 15B) have been available from all parts of the Altaid collage and from the Kuen-Lun system at large (including the Qilian Shan, Qinghai Nan Shan, and the Anyemaqen Shan) ever since Suess (1901), Kober (1921), Argand (1924), Lee (1928), and Stille (1929) drew attention to its widespread occurrence and especially after interest in it was revived in the framework of plate tectonics by the landmark paper of Molnar and Tapponnier (1975). We here cite only a few of the more comprehensive, exemplary, and the most recent studies documenting it: Şengör (1984; 1987), Hendrix et al. (1992), Lu et al. (1994), Allen et al. (1995), Cunningham et al. (1996), Allen and Vincent (1997), Burov and Molnar (1998), Delville et al. (2001), Greene et al. (2001), Ritts and Biffi (2001), Sobel et al. (2001), Sjostrom et al. (2001), Vincent and Allen (2001), Chen et al. (2002), Wartes et al. (2002), and Xiao et al. (2004).

Paleomagnetic measurements bear out our estimate. Our depiction of the position of north China plus the Manchurides in Figure 15B, based on geological estimates, is the position found appropriate on the basis of the available paleomagnetic observations by Enkin et al. (1992), which are corroborated by newer observations in various points in Siberia (e.g., Kravchinsky et al., 2002a, 2002b), and since used in many reconstructions such as those by Scotese (2001 and at http://www.scotese.com/newpage5.htm, visited on 16 December 2007).

However, Gilder et al. (2008) deny the large amounts of shortening within Eurasia north of the Tarim basin simply because they think no ocean closed there after the Paleozoic. This is untrue for almost the whole of Mongolia. Şengör and Natal'in (1996) documented the evidence showing the closure of the Mongol-Okhotsk Ocean as a consequence of the pair of scissors-like closing of the two legs of the Tuva-Mongol fragment. Farther west there has been much intracontinental shortening as mentioned above.

However, there has also been considerable Mesozoic and Cenozoic deformation within the north China block itself and along its borders, along the Helan Shan, Yinshan, and the Taihang Shan (southern part of Luliang Shan) trends, known since the great pioneering studies by Teilhard de Chardin and Licent

(1924), and which is now being gradually unraveled, together with the deformation of the northeasternmost parts of the Tibetan Plateau, among many others, especially by the painstaking field studies of Clark Burchfiel and Greg Davis and their associates (see, e.g., Zhang et al., 1990; Burchfiel et al., 1991, 1992b; Darby et al., 2001, 2004; Davis et al., 2001, and the references there, 2002, 2004; Davis, 2003; Liu et al., 2005). Davis (2003), in particular, drew attention to the significance of the adakitic magmatism marking the long interval (from ~170 to 130 Ma ago) of shortening within the north China block as a consequence of the Panthalassan-Pacific subduction synchronously with the last episodes of the Altaid-related shortening in Mongolia and the Russian Far East (cf. Şengör and Natal'in, 1996). Şengör and Natal'in (1996) have palinspastically restored the north China block to its early Triassic shape using all the observations available to them prior to 1995 (their fig. 21.53F), and it is their result we adopt here pending the synthesis by Greg Davis.

To the present north of the Manchurides, the Tuva-Mongol fragment completed its collision with the Manchurides also in the latest Permian along the Solonker suture (Fig. 15B; Şengör and Natal'in, 1996; Xiao et al., 2003).

In contrast to north China, south China consisted of two main pieces separated by an ocean in the late Permian: The Yangtze (Y in Fig. 15B) and the Huanan (H in Fig. 15B) blocks separated by what Şengör et al. (1988) called the Xiangganzhe suture zone of latest Permian-Triassic ages (Şengör et al., 1988; Hsü et al., 1988, 1990). Almost all the high quality paleomagnetic data come from the Yangtze block and show it to straddle the late Permian equator with less than 20° latitudinal and less than 25° rotational (around a vertical axis) uncertainty (see Enkin et al., 1992, fig. 22). This is perfectly compatible with the geology of its borders. It has long been shown that it contacted the north China block in the late Permian (Okay et al., 1989, 1993; Ernst et al., 2007), and the suture closed diachronously from east to west, creating an immense flysch wedge of early Triassic age that formed the "Yajiang Flysch" of the Songpan Ganzi system (Şengör and Natal'in, 1996; for sedimentological details and the stratigraphic subdivisions of the "Yajiang Flysch," see Nie et al., 1994[17]). This suggests that the Yangtze block rotated by almost 90° with respect to the north China block (but *not* with respect to the spin axis of Earth) since the late Permian and thus zipped close the Qin Ling suture by late Triassic time (see especially Zhou and Graham, 1996).

The position of the Huanan block is constrained paleomagnetically to lie some 10° farther south than the Yangtze block in the early Triassic in an orientation parallel with that of the Yangtze block (H.H. Chen et al., 1992, 1993). We note this, but also underline that at such small separations, the method is not always terribly reliable. We know from its paleobotanical affinities to the Yangtze block, however, that the Huanan block was in contact with the Yangtze block already in the early Permian (Fig. 15D1), as also indicated by its late Permian benthic foraminifera (Fig. 15F; Şengör et al., 1988). Yet the Xiangganzhe Ocean did *not* close until late Triassic time, because between the

Yangtze and the Huanan blocks, the so-called Banxi complex[18] includes Permian and Triassic pillow lavas, radiolarian cherts, and deep-sea turbidites; in some of the pelagic Banxi rocks, typical early Triassic Tethyan ammonite genera such as *Otoceras* and *Ophiceras* have been found showing that the final closure of the Xiangganzhe suture must have post-dated their deposition (Şengör and Natal'in, 1996; Xiao and He, 2005). The straightness of the suture (that also localized later strike-slip), its paucity of arc magmatic rocks (cf. Şengör, 1991b), and the *en échelon* nature of the nappes that have resulted from its closure, to us suggest that the suture was of a type that Dewey et al. (1986, fig. 18) called a "transform suture." We infer that the Huanan block slipped into place along a highly oblique convergent zone between the early Permian and the late Triassic.

Until the late Permian, the Yangtze block had been fringed by a continental margin arc to its west and southwest. That arc began rifting from it in the late Permian concurrently with the eruption of the Emei Shan basalts (cf. Courtillot et al., 1999; in Fig. 15B, the isotopic age 259—Ma—on the Yangtze block indicates their location) and opened behind it the Yajiang marginal basin (Ym in Fig. 15B). The resulting ensialic island arc has been called the Shaluli Shan arc (Şengör, 1984; see esp. Görür and Şengör, 1992, fig. 2; Sh in Fig. 15B), and it is now located to the west of the Songpan-Ganzi system forming the outermost of the three magmatic arcs delimiting the eastern Qangtang block (Şengör, 1984; Şengör et al., 1988; Şengör and Natal'in, 1996). The eastern Qangtang (EQ in Fig. 15B) block in Tibet has late Permian marginal marine to shelf sedimentary rocks consisting of carbonaceous shales intercalated with fossiliferous calcareous sandstones, dark gray thinly bedded limestones, coal deposits, and basalts, andesites, and trachyandesites. The volcanics overlie essentially Capitanian sedimentary rocks and are correlated with the Emei Shan basalts dated to be ~259 Ma old (Ali et al., 2005). On the face of it, they seem at least 1 Ma *younger* than the Guadalupian-Lopingian boundary as accepted by the International Geologic Time Scale (Gradstein et al., 2004). Ali et al. (2002) showed, however, that they are at

---

[17]For the details of the sedimentology of the flysch deposited outside the Yajiang marginal basin, south of the present-day Anyemaqen Shan, see Zhou and Graham (1996).

[18]Banxi is a basket term for anything that has ophiolites, mafic volcanics, and deep-sea clastics and cherts commonly in a mélange. Some parts of it are undoubtedly Precambrian, others undoubtedly Permo-Triassic. Its equivalents in different provinces are designated as follows: Xiajiang Group in Guangxi and Guizhou, Shangxi Group in Anhui, Shuangxiwu Group in Zhejiang, and Jianou Group in Fujian. Work is now underway to separate different rock groups belonging to different ages and to different tectonic contexts, but the outcrop conditions dictated by the humid climate of south China make it difficult. Studies such as that by Liu et al. (1996) deny the Triassic suturing here, because they consider only the Precambrian part of the old Banxi Group *sensu lato* and ignore the younger ages known since the older studies by Yang et al. (1981). In the field, Şengör himself was shown the Wan Zhe Gan nappe in Anhui in September 1986, for example, which is bordered by an ophiolitic mélange of late Paleozoic to early Triassic age overridden from the southeast by the Wan Zhe Gan nappe. All of these are considered part of the Shangxi Group (Banxi)! Indeed, the interpretation first proposed by Ken Hsü concerning the existence of the Xiangganzhe Ocean and its early Mesozoic obliteration has recently been corroborated by Xiao and He (2005). They chose to call the vanished ocean the Jiangshan-Shaoxing deep sea, however.

least two conodont biochronozones, i.e., 1–1.5 Ma *older* than the Capitanian-Wuchiapingian boundary. They pointed out, however, that the overlying reverse-polarity flows and the explosive volcanic waning sequence could be coeval with the Capitanian-Wuchiapingian boundary. The Emei Shan volcanic rocks seem related to the rifting event that separated the Shaluli Shan arc from the Yangtze block and may record the opening of a slab window because of ridge subduction (see below).

The position of southeast Asia during the late Permian is a complicated issue because it is still unclear how many pieces at the time made it up. One clearly delineated entity is Annamia (Şengör, 1984; Şengör et al., 1988; Şengör and Natal'in, 1996; A on Fig. 15B), i.e., the nuclear Indochina (Metcalfe, 1996, fig. 2), called Indosinia by Hutchison (1989, p. 106–107). By the late Permian, the nuclear cratonic piece of Pan-African age had already acquired quite a halo of Paleozoic orogenic belts around it. Where it was precisely at the time is unclear because its pre-Triassic paleomagnetic record was obliterated by a block-wide remagnetization (Chen and Courtillot, 1989; Richter and Fuller, 1996). However, it shares a south Cathaysian flora with the Yangtze and the Huanan blocks, has the very same benthic foraminiferal assemblage (Şengör et al., 1988), and has many *Dicynodon* fossils (Battail et al., 1995, 1997, 2000; Steyer, 2008, and 2007, personal commun. via Şevket Şen; see also Buffetaut, 1989) indicating land connections with Gondwana-Land and possibly also with Laurasia. It also has a part of the late Permian Emei Shan basalts on its (present) northeast edge (Ali et al., 2005; the small volcanic symbol Λ, labeled El on Annamia in Fig. 15B). It is therefore clear that, at the latest by the medial Permian, it had established contact with the south Chinese assemblage. Before that, paleobiogeographic data show it to have been a part of northern Gondwana-Land (e.g., Burrett and Strait, 1986), most likely in contiguity with the Yangtze block with which it shares very primitive Yunnanolepiform antiarchs that occur nowhere else (Thanh et al., 1996). Most authors agree that the contact of Annamia with the south Chinese ensemble had to have occurred between the Huanan and the present east or even southeast side of the Annamia block, about where the present-day Da Lat massif, located some 200 km to the northeast of Ho Chi Minh in Vietnam, where late Paleozoic orogenic deformation has long been known to have occurred (Hutchison, 1989). We believe that the Emei Shan equivalents in Laos were later strike-slipped to their present locations from here.

The Song Da suture (Şengör, 1984; Şengör et al., 1988; Şengör and Natal'in, 1996), to the present northeast of Annamia, had remained open until the early Triassic as has recently been corroborated again by the oblique-collision-related, suture-parallel dextral shear zones dated to the end of the Scythian (Lepvrier et al., 1997). Clearly, Annamia rotated rapidly clockwise after its (present-day) southeast corner hit Huanan and, after it had completed its rotation, began slipping along its newly completed suture zone.

We do not distinguish an east Malaya block (Metcalfe, 1996) from the rest of Annamia; even Metcalfe himself treats

them together. It is most likely that the southwest Borneo block (SWB in Fig. 15B) was also a part of Annamia, but, again, we have no evidence one way or the other. Let us say that there is no reason to separate them during the late Permian. Following Metcalfe (1996) we treat it as part of Annamia.

The western part of Thailand and Malaysia, comprising easternmost Burma, has been variously termed Sinoburmania or Sinoburmalaya (Hutchison, 1989), Shan-Thai (Bunopas and Vella, 1983), or Sibumasu (Metcalfe, 1984; S in Fig. 15B), but it is nothing more than the easternmost part of the Cimmerian continent that began rifting in the Tournaisian from the Mount Victoria Land block (now in Burma: Mitchell, 1989; MV in Fig. 15B) that had as yet remained attached to Australia (Görür and Şengör, 1992, fig. 5). Sibumasu was attached to the west Changtang block (WQ in Fig. 15B) via the Bao Shan block of easternmost Tibet (too small to be shown in Fig. 15B: Şengör et al., 1988; Şengör, 1993; Şengör and Natal'in, 1996; Akçiz et al., 2008), although the character of the basement of the west Qangtang block is not known: from the few reliable descriptions of Norin (1946, 1976, 1979) it looks as if at least its northwestern section may be a late Paleozoic subduction-accretion complex in which the distal parts of the Carboniferous-early Permian Horpa-Tso and the Permian Tashliq-Kol clastic sedimentary rocks may have been involved.

Even the character of the basement of the Sibumasu part of the Cimmerian continent is not well known, largely because of the extensive jungle cover in Thailand and Burma. However, Şengör and Natal'in (1996) show much of the basement of the Sukhotai foldbelt as Permian-Triassic subduction-accretion complex owing to the widespread occurrence of graywackes and greenschist metamorphic rocks. Whatever its nature, by Permian time it was basement to a (now) east-facing magmatic arc forming the continental margin orogen of the Sukhotai foldbelt (Şengör, 1986, fig. 1b). The Chiang Mai ophiolites (Ricou, 1994, 1996) and pelagic sedimentary rocks including radiolarites (Wonganan and Caridroit, 2006; Randon et al., 2007) may well represent a marginal basin that opened behind this magmatic arc, but closed later than it, much like the still open Black Sea marginal basin behind the already collided Rhodope-Pontide arc in Turkey (Okay et al., 1994).

The ocean that closed between the Sukhotai and the Loei foldbelts gave rise to what is in the literature known as the Nan-Uttaradit–Sra-Kaeo–Bentong-Raub suture (Şengör et al., 1988; Şengör and Natal'in, 1996; Metcalfe, 1996) and Şengör (1986) showed, in some local detail from northern Thailand, that it did not close until the late Triassic, although oceanic pelagic sedimentation in it had ceased already in the Permian and the narrowing strait was invaded by the thick flysch deposits of the Lampang Group (see Şengör, 1986, fig. 2). *As we shall see below, this somewhat sediment-choked strait, together with the Chiang Mai basin, was the only relatively deep marine connection of the Paleo-Tethys with the Panthalassa, and it may not have had abyssal depths at all despite its two flanking subduction zones, much like the present-day Molucca Sea (see Şengör, 1990d, figs. 10N and 10O).*

The early Carboniferous rifting in northwest Australia appears to have propagated westward along two different lines of tectonism: by late Carboniferous-early Permian time it had reached the Gondwana rifts in India; the Damodar-Koel Valley, the Son-Mahanadi, and the Kurduvadi rifts (DK, SM, and K, respectively, in Fig. 15B) commenced their activity in the late Carboniferous (Şengör and Natal'in, 2001). This line could not have gone through the western coastlands taphrogen in Australia (WC in Fig. 15B) because after an initial phase in the Ordovician that system of extensional structures were not revived until the Permian (Şengör and Natal'in, 2001). However, in the Lhasa block, some 25 km east of Lhasa (just about where the letters BO stand in Fig. 15B), the Dagze basalt-andesite sheet flows and tuffs of late Carboniferous age with a thickness of some 1500 m indicate a subsiding environment in a magmatic arc (Pearce and Mei, 1988). This, to us, suggests rifting near the axis of a north-facing subduction-related arc (as there could not have been any arcs to the south in a cratonic area!) and continues the early Carboniferous setting similar to that in Thailand and northwest Australia westward. It is likely that once the upper plate—in this case the Gondwana area in India—was put into extension, the typical Gondwana rifts of India much farther south may also have formed. The fact is, however, that rifting, somehow, progressed from the east (in Pangean coordinates), from Thailand and northwest Australia, westward toward India and Tibet.

The other line of rifting west of Thailand was not far from the rifts we just mentioned. It was farther north and essentially rifted the west Qangtang (WQ in Fig. 15B)–central Pamir–Farah strip (F in Fig. 15B) from the Lhasa–southwest-southeast Pamir–Helmand blocks (as yet integral parts of Gondwana-Land) along the Bangong Co-Nu Jiang–Rushan-Pshart–Wašer oceanic area (BO in Fig. 15B; Şengör et al., 1988; Şengör and Natal'in, 1996). In eastern Tibet, in the Qamdo Prefecture east of the Lancang River, this rifting episode is recorded by the nearly 2-km-thick Lower Carboniferous Machala Formation, consisting of thickly bedded limestones intercalated with silty mudstones, tuffs, arkosic sandstones, and schistose andesites and basalts (Yang et al., 1986; Fan, 2000). It seems as if the Yunzhu Group on the Lhasa block side consisting of Lower Carboniferous sandstones interbedded with shales, limestones, and marls is the other side of the now split rift; in the whole area, vulcanicity continued into the Permian (Chang et al., 1989). The rifting began later, in the Permian, in the Rushan-Pshart and, earlier again, in the early Carboniferous (Tournaisian!) farther west near the former Wašer Ocean (WO in Fig. 15B; for age of volcanics in the Taš Küprük unit in Afghanistan, see Gaetani et al., 1996, p. 697) and was also accompanied by vulcanicity[19] (Şengör and Natal'in, 1996). In his lavishly illustrated farewell lecture following his retirement from active teaching, the learned and indefatigable Tethyan master Maurizio Gaetani reproduced his map from Gaetani (1997a, fig. 9, and 1997b, fig. 6) giving an excellent summary of the geometry of the Permian rifting in the Karakorum, the Pamirs, and westernmost Tibet (Gaetani, undated [2006], map on p. 18). There, it is clear that some of the black slates of western Qang-

tang, the Pamirs, and the Karakorum[20] (Norin, 1976, 1979) must have been laid down in these nascent oceans that communicated directly with the Paleo-Tethys.

This communication was tied to the Sistan Ocean in eastern Iran (Şengör, 1990c) and from there directly to the nascent Neo-Tethyan rift in Oman, where, by early late Permian time, radiolarite deposition had already commenced on the distal apron of a northeast-facing continental margin (Blendinger et al., 1990, fig. 7) along with ongoing Hawasina mafic magmatism that resembles the Plio-Pleistocene marginal basin magmatism of the Tyrrhenian Sea (Şengör, 1990c).

Farther northwest, in the Zagros, rifting occurred later (ZR in Fig. 15B). Most sequences there show rapid, rift-related subsidence, commencing at the beginning of the Wuchiapingian, but the Upper Permian Dalan Formation was still affected by normal faulting (Szabo and Kheradpir, 1978). This Permian rifting seems to have characterized the entire periphery of the eastern Mediterranean, except its western margin in the Malta Escarpment and Italy, which appear to have rifted later. Late Permian rifting also characterized the entire Apulian Shelf (the entire light gray area marked AS in Fig. 15B) from Tunisia (Solignac and Berkaloff, 1934; Busson, 1969, Ben Ferjani et al., 1990, p. 33–35), via Sicily, to the eastern Alps and the basement of the Pannonian basin.

Beyond the Pannonian basement, the northernmost part of the Apulian Shelf in the world of the late Permian, we come to the end of the regions where Alpine-type Triassic is encountered, namely to North Dobrudja. In this place, the non- to little-metamorphic pre-Triassic sequence had long been shown to consist of the enigmatic Carapelit Formation. The enigma concerning the Carapelit stems from the fact that no agreement has so far been reached either about its age or about its milieu of deposition. Published estimates range from top Devonian to Permian and interpretations of depositional environments from alluvial fan to marine flysch (see, for example, Seghedi and Oaie, 1986)! This is probably because one has lumped under the designation "Carapelit" widely different associations that contain all of the above. We here cite Şengör's personal observations made in April 1992 under the guidance of Mircea Sandulescu and Eugen Gradinaru in the southernmost parts of the Carapelit outcrops just west of the town of Camena, north of the well-known ore locality of Altın Tepe ("gold hill"), where a southerly dipping succession exposes a part of what is here also mapped as Carapelit (Fig. 16). Here a greenish to reddish silty-sandy turbidite sequence commences atop the deformed and metamorphosed Paleozoic rocks, and it is dated as late Permian to early Triassic. Above are early Triassic rhyolite porphyries and a small thickness of Ladinian-Lower Carnian limestone followed by further rhyolites.

---

[19]It is to us actually unclear whether the Taš Küprük volcanics are products of a rift or a continental margin magmatic arc. They are said to be mainly porphyritic basalts and tuffs interbedded with marine limestones of Tournaisian age (Gaetani et al., 1996, p. 697). The situation may be very similar to the case of the Yunzhu Group in eastern Tibet mentioned above.

[20]Lumped into the Wakhan (Hayden, 1915) and the Misgar (Desio, 1963) slates as indicated by Gaetani et al. (1995, 1996) and Gaetani (1997a).

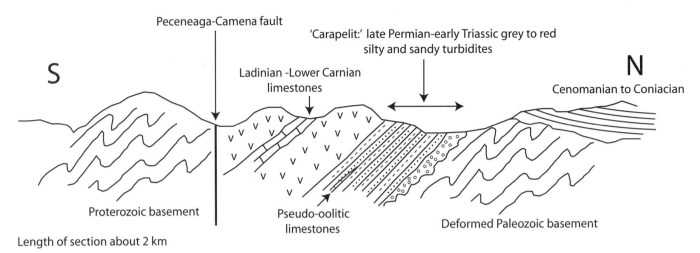

Figure 16. North-south simplified cross-section near Camena, Romania, drawn on the basis of Şengör's field notes taken on 13 April 1992. This is just to the northeast of the southern border fault of the rift labeled DP in Figure 15B.

In the *locus typicus* of the Carapelit, located in a syncline north of Mecina, however, the formation consists of a red breccia with angular clasts of red sandstones and granites, felsic volcanics, red sandstones, and quartz. In fact, at first sight, one gets the impression of standing before a typical Verrucano outcrop. It is in this sort of facies that Oaie (1986) documented an alluvial fan-river flood-plain deposit laid down during an episode of active faulting. In such areas, the sedimentary rocks in places are interbedded with volcaniclastics. Elsewhere, Şengör was told by his guides that Carapelit showed rythmic sedimentation. The Carapelit is clearly different things in different places, but its sequence makes sense: the picture that gradually emerges of it is of a rift sequence that began forming sometime in the late Carboniferous, depositing terrestrial sequences and volcaniclastics. The sequence must have turned marine at the latest by Ufimian (late Kungurian) time, because we know that marine faunas from the Paleo-Tethys at this time invaded the Permian marine basins in Silesia (Peryt and Peryt, 1977) and farther

north the "Zechstein basin" in northern Germany (as known from the bivalve *Bakevellia ceratophaga* Schlotheim: Merla, 1930), and the Măčin unit in the North Dobrudja, in which the Carapelit rift was located, was the only avenue that we know of that could have been used. It thus seems that not all evidence of marine connection of the Paleo-Tethys with the marine basins in the north had since been eroded as Ziegler (1990, p. 74) assumed and, in fact, *Bakevellia ceratophaga* common to the Lower Saxony basin in northern Germany, to the top Permian–early Triassic Gerennavár Formation of the Bükk Mountains in Hungary and to the latest Permian *Bellerophon* Formation of the Southern Alps shows that the marine connection had been maintained to the end of the Permian and beyond. This is also shown rightly by Ziegler et al. (1997) in their Tatarian (late late Permian) reconstruction.

A continuous chain of rifts and post-rift basins connected the Carapelit rift with the Norwegian/Greenland rift system in the far north of the Pangean world (Figs. 15B and 15C, 17A and

Figure 17. (A) Late Permian tectonics and paleogeography of Europe and surrounding areas showing faults and volcanics dating back to the latest Carboniferous (see the various papers in Wilson et al., 2004). Lines with hachures are normal faults with hachures on the hanging wall. Red upside-down V's are rift-related volcanic rocks. The yellow arrows show possible Paleo-Tethys to Pangea water transfer direction documented by faunal transfer as far as central Europe. The rest is inferred by analogy to water transfer from the Atlantic to the Red Sea during the Messinian salinity crisis (Ryan, 2009). This was most likely intermittent and dependent on the sea level in the northern and northwestern European basins. Often the flow was reversed. For sources, see: rifts: Wopfner (1984), Betz et al. (1987), Ziegler (1988, 1990); magmatism: main database is Wilson et al. (2004), supplemented by Sanchez Carretero et al. (1990), Zheng et al. (1991–1992), Rottura et al. (1998), and Glennie et al. (2003). The offsets of Hercynian fronts are from Franke (2000) and Rossi et al. (2008) for the southern front and Ziegler (1988, 1990) for the northern front. Shoreline is from Ziegler (1988, 1990). Most Polish authors depict the Hercynian front as *not* laterally offset (e.g., Dadlez, 2006; Królikowski, 2006), but we think this unlikely, not only because the Hercynian and post-Hercynian structures interfere even on their own maps (e.g., Dadlez, fig. 1.A), but also because many of the geophysical anomalies are far too angular on a large scale indicating truncation and not gentle curvature of the front (e.g., Królikowski, 2006, figs. 2 and 3). The Bohemo-Sardinian shear is depicted broadly equivalent to the external crystalline massifs shear zone (ECM-SZ) (see Rossi et al., 2008, p. 49, map). (B) A schematic cross section from the Dobrudja Porte to the Norwegian-Greenland Sea rift system showing the gradual rise of the top surface of anoxic waters in the Paleo-Tethys and their spilling into the European–future North Atlantic regions. The superb synthesis by Littke et al. (2008), describing the various aspects and the evolution of the Lower Saxonian–North Sea Permian Basin, arrived in Istanbul too late to be used in this study, but its contents corrobarate what is said here.

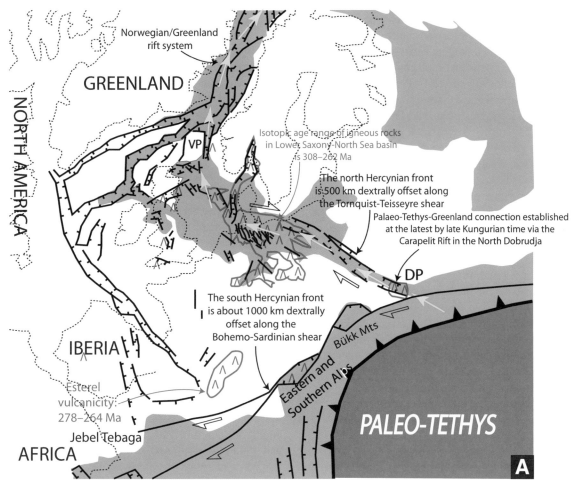

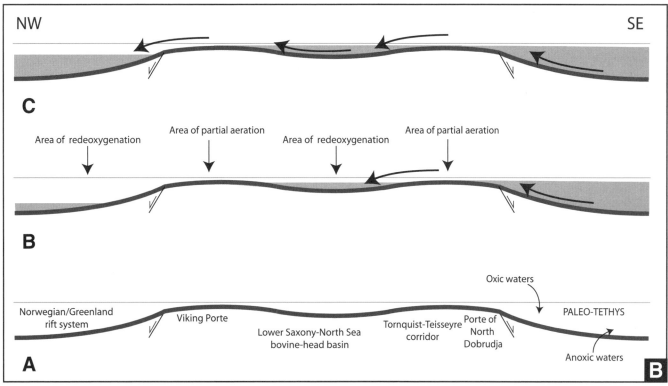

17B). From the kinematics of the Permian tectonics, it seems clear that the right-lateral motion along the Scythides (Natal'in and Şengör, 2005) was separated into two strands west of the present-day Black Sea: one strand went toward the west-north-west, only roughly following the early Paleozoic Thor suture (Berthelsen, 1998) and creating a transtensional system along the Carapelit–Tornquist-Teisseyre lineament (Figs. 15B and 17A; also see Dadlez, 2006, fig. 1.A for the *en échelon* arrangement of the main faults along it indicating right-lateral displacement along the zone). This lineament bounded on the east the Lower Saxony basin (see Dadlez, 2006, for a detailed structural picture of that contact), consisting of a closely spaced array of mainly extensional structures resembling tension gashes along developing shear zones. This extensional structure was established in the early Permian and was coeval with a widespread Permian rift-related magmatism (Fig. 17A; Betz et al., 1987; Ziegler, 1988, 1990; Glennie et al., 2003; Timmerman, 2004; Dadlez, 2006). In the late Permian, a bovine-head-shaped basin formed atop the deactivated normal faults (van Wees et al., 2000). The same was true in the North Sea basin to its northwest. These shallow basins became sites of extensive evaporite deposition during the Zechstein, but both carbonates and clastics were also continuously deposited throughout the Zechstein time in different parts of both the Lower Saxony basin and the North Sea basin.

It had long been assumed that both these basins had been fed exclusively by the Norwegian-Greenland rift corridor that allowed Arctic waters to come into the heart of Europe to pump the evaporation pans. We have seen, however, that this was not exclusively the case. The Zechstein basins of Europe were fed both from the north *and* from the south. It is impossible to tell which way the dominant current was; it was probably different at different times, but in the late Permian, anoxia from the North Dobrudja to Greenland developed from south to north! We take this as evidence that at least when anoxia moved north, the current was from the Paleo-Tethys to the Norwegian-Greenland rift corridor. Whether waters from one ocean can traverse such long distances across evaporative epi- and intracontinental basins to reach another ocean is answered in the affirmative by the arrival of the Atlantic waters to the Red Sea during the Messinian salinity crisis (Ryan, 2009). It was thus not impossible for Paleo-Tethyan anoxic waters to reach the Greenland/Norwegian rifts along the European rift corridor.

The question then becomes what business anoxia had within the late Permian Paleo-Tethys? Was it really there? How and when did it develop? If it existed, how extensive was it?

Before we answer these questions we must try to discover what the bathymetry of the Paleo-Tethys might have been like because basin configuration is one factor that strongly influences aeration in the seas. Unfortunately, no part of the oceanic floor of the Paleo-Tethys survives and all of its continental margins, including its shelves, have suffered subsequent orogenic deformation not only during its own closure, but also during the obliteration of its successor, the Neo-Tethys (Şengör, 1984, 1987; Şengör et al., 1988; Şengör and Natal'in, 1996). The only way

we can go about finding out what its floor was like is by looking at the plate kinematics at the time. Here, an additional unfortunate aspect is that the Paleo-Tethys was completely surrounded by subduction zones, making it impossible to link, in a rigorous way, any part of its interior plate kinematics to its surroundings. But, still, there is a lot one can do.

First, we establish the orientation of the margins of the Paleo-Tethys and their character. What their character was we have already mentioned: they were all Pacific-type, i.e., subduction margins. Their orientations are given by the reconstruction shown in Figures 15B and 15C. If the Pangea were a single plate, the condition that all Paleo-Tethyan margins be subduction zones could only have been fulfilled, provided our reconstruction of the Paleo-Tethyan margins are correct, if at least three ridges existed in it. Let us start with the single plate assumption, knowing it not to be true. What enables us to assume it anyway is that there were no obvious areas around the Paleo-Tethys moving toward it with velocities similar to those of typical subduction zones independently of the rest of the Pangea (all the circum-Paleo-Tethyan marginal basins had just begun rifting and, as such, were in the very slow phase of their development).

The next thing we know is that the northern margin of the Paleo-Tethys had a very strong dextral strike-slip component at this time (Natal'in and Şengör, 2005). There the subduction could not have been head-on, but must have been oblique toward the west to northwest in Pangean coordinates. These conditions could be satisfied by a ridge geometry shown as P and p in Figures 15B and 15C.

Most marginal basins of the Paleo-Tethys were along its Gondwanian margin. In fact, that whole margin was predominantly extensional during much of late Paleozoic time. By analogy with the present-day Pacific Ocean, we thought it appropriate to put the oldest Paleo-Tethyan oceanic crust in front of those basins that were then facing to the east-northeast. That placement pushed the northwest-trending spreading center toward the Cathaysian bridge. If the Emei Shan basalts are to be related to some sort of a plume (shallow or deep), one would need a slab window there. A ridge subduction is the ideal geometry in which to generate one. If the ridge R (Figs. 15B and 15C) went down the subduction zone before the eruption of the Emei Shan basalts that would do it, and it is consistent with what we have so far said. In our reconstruction we did *not* show such a ridge subduction (it should have been shown) only to be able to illustrate what the geometry of the ridge might have been like before it became consumed.

A particular problem has always been the absence of voluminous arc magmatism in northern Turkey and Iran during the Paleo-Tethyan subduction. If the ridge P really was being subducted in the manner shown, then this explains not only the lack of arc magmatism, but uplift associated with the consumption of a spreading center would also explain the near-absence of Permian sedimentary rocks from northernmost Turkey and the abundance of mafic magmatism in the back-arc area in the future Karakaya basin (Şengör and Yılmaz, 1981). The early late Permian (260 Ma ago) low-grade metamorphism in the eastern Rhodope-Pontide fragment discovered by Topuz et al. (2004) corroborates our inference of the ridge subduction here.

The way we depicted the ridges generates plate motion in such a way that a ridge could function disrupting a subduction zone in the Pamirs as shown (ridge east-northeast of the letter F in Fig. 15B[21]).

In short, the ridge geometry presented in Figures 15B and 15C does a remarkable job in explaining the circum-Paleo-Tethyan tectonics in the late Permian. A by-product of it is the bathymetry of the Paleo-Tethys. As ocean floor subsides away from the ridges following a law dependent on the square root of its age, we can precisely draw the bathymetry of the Paleo-Tethys once we know the distribution of the ridges within it. Figure 15C shows this plus the topography of the Pangea mostly after Ziegler et al. (1997).

The depth ranges of the Paleo-Tethyan abyssal basins must have been prodigious because they must have gone from -2.5 km[22] at the ridges to trench depths in less than 3000 km. This strong topographic differentiation probably helped to inhibit basin-wide circulation in the Paleo-Tethys in a way suggested by Tissot et al. (1979) for the early Cretaceous eastern central north Atlantic.

As a final support of our reconstruction of the Pangea, Figures 15D1 through 15F show some of the latest Carboniferous to earliest Triassic biospheric elements on it. The maps are self-explanatory, but certain salient aspects may require some comment: In Figure 15D1, one sees the strong dependence of the floral realms on latitude and hence climate. This is an aspect often noted (e.g., Ziegler, 1990) and has its parallels in the fauna (Sidor et al., 2005). The climate is not only latitudinally controlled, but also by proximity to certain seas: in north China, for example, the warmth-loving north Cathaysian flora developed closer to the warm Paleo-Tethys, whereas the cold-loving Angara flora developed closer to the Panthalassa, although they overlap in latitude considerably and get mixed precisely in these very areas, as Sun (2006) recently emphasized again. The mixed Cathaysia-Angara floras are restricted to some areas in north China, northwest China, and to Primorye (Russian Far East). These mixed Cathaysia-Angara floras are dominated by Angaran elements of late Permian age. The Cathaysian and Angaran floras began to merge with one other because of the collision of the north China and the Tarim blocks with the Altaids, and, we think, because they overlapped latitudinally along the shores of two oceans of contrasting climatic regimes. The floras become more mixed as their representatives are farther removed from the shores of these oceans. Similarly, the flora of the Southwestern United States is distinct from the colatitudinal Euroamerian flora; the

former has swamp-marsh loving plants, whereas the latter is adapted to dryer conditions.

The south Cathaysian flora (red squares in Fig. 15D1) occupies the Cathaysian bridge entirely, although representatives of it are found as far west as southeastern Turkey showing that at least at times the bridge was complete.

The distribution of the genus *Lystrosaurus* of the latest Permian to late Olenekian time (Fig. 15D2) is most instructive. This sheep-size dicynodont originated in east Africa in the latest Permian (Cosgriff, 1965; King and Jenkins, 1997; Lucas, 2006) and rapidly colonized the eastern half of the earliest Triassic Pangea, using the Cathaysian bridge to cross over to Laurasia going as far west as the Moscow basin, but not farther into the arid world of western Pangea. It has commonly been viewed as a cosmopolitan animal with universal distribution in Pangea. The reality, however, is not so. *Lystrosaurus* is entirely absent from North America and western and central Europe and no corroborated occurrence has yet been reported from South America. Although Benton (2004, p. 16) writes that *Lystrosaurus* also occurs in South America, we have not been able to corroborate that the stapes reported from the Sango do Cabral Formation in Brazil actually belongs to a *Lystrosaurus* (Spencer G. Lucas, 2007, written commun. via Bruce Rubidge).

The *Lystrosaurus* finds on the southern part of the Cathaysian bridge, including a possible instance from Australia, are also disputed. The 1923 report on a *Lystrosaurus* from Luang Prabang in Laos from the Annamia side of the Nan-Uttaradit suture (cf. Buffetaut, 1984, fig. 1), found in the 1890s by Jean-Baptiste-Henri Counillon (1896), supposedly in the basal horizon of the Pou Say sandstone (then thought to be entirely early Triassic, because all *Dicynodon*s were then believed to be Triassic: see Steyer, 2009), and first described by Joseph Répelin in 1923, has not been confirmed by any subsequent detailed study because the specimen can no longer be located, although Colbert (1975) accepted it as *Lystrosaurus* without saying why. Buffetaut (1984) gives a brief history of the controversy concerning the true identity of that specimen and its causes and Battail et al. (1995) present another summary, correcting the age of the rocks bearing the fossils as Tatarian, i.e., uppermost Permian. Steyer (2009) recently wrote the most comprehensive historical narrative, with an exhaustive literature survey, of the Pavie Mission (1879–1895), of which Counillon was a part. King (1988, p. 96–97) lists the Luang Prabang find by Counillon as *Lystrosaurus incisivum*, despite Buffetaut's hesitations. However, the careful critical assessment and the figure Battail (1997, fig. 2) published now, we think definitively, removes any doubt and resuscitates Piveteau's (1923) assignment of the now-lost skull fragment to the genus *Dicynodon*. In fact, additional *Dicynodon* fossils have more recently been collected from the same locality (see Steyer, 2009). That is why we do not show any *Lystrosaurus* in Indochina in our Figure 15D1.

Despite this negative result, we cannot imagine how *Lystrosaurus* could have reached north China and central Asia without going through the Cathaysian bridge because it does not occur

---

[21]Naturally the actual geometry of this disruption was more complicated as documented by Gaetani (1997a, 1997b), but Gaetani's reconstruction is perfectly compatible with the requirements of the kinematics we deduce here.

[22]Probably a couple of hundred meters deeper, because by late Permian time no ice caps of any substantial dimension had been left on Earth.

anywhere in western Pangea, despite the presence of sedimentary deposits of appropriate age and facies that yielded tetrapod fossils (Fig. 15E).[23] The migration of the *Lystrosauri* during the latest Permian and the earliest Triassic thus must have taken place along the Cathaysian bridge and provides corroboration that by that time it had to have isolated the Paleo-Tethys from the Panthalassa. In fact, more than a decade ago, our meticulous and critical friend Bernard Battail (1997) independently reached the conclusion that a land bridge between Laurasia and Gondwana-Land must have existed to the *east* of the Paleo-Tethys on the basis of the distribution of the genus *Dicynodon* (also B. Battail, 2007, personal commun.), known since the beginning of the twentieth century from Russia (Amalitzky, 1901). Colbert in 1972 already had pointed out that the Chinese (and by implication the Russian) *Lystrosauri* must have gone there directly from Gondwana-Land (Colbert, 1975).

The studies by Tazawa and Chen (2005) on the brachiopods of the Tumenling Formation (Wordian-Capitanian) in the Wuchang area in the southern Heilongjiang Province of northern China document the unity of the northern part of the bridge already in the later medial Permian time. The Wuchang brachiopod fauna, as a whole, is found by them to be comparable to several contemporaneous faunas described from Inner Mongolia in northern China, Jilin in northeast China, south Primorye in eastern Russia, and the Hida Gaien belt of Japan (i.e., the old, pre-Permian nucleus of Japan: see Şengör and Natal'in, 1996, for its paleogeographic relation to the other areas here mentioned), in terms of its index elements *and its boreal–paleoequatorial mixture*. During the medial Permian these areas were referred to as the Inner Mongolia–Japan transition zone or the northern transition zone between the paleoequatorial and boreal realms. The overwhelming majority of the boreal elements reveal that the Wuchang area was probably situated in the northern subzone of this transitional zone, which belongs to the Manchurides, i.e., to the north China block. *These east and northeast Asian blocks acted as migratory stepping stones providing a bridge for faunal migration between the paleoequatorial and boreal realms during the medial Permian* (Tazawa and Chen, 2005).

The distribution of the benthic foraminifera of the late Permian in the Paleo-Tethys, displayed in Figure 15F, show yet again the same thing, i.e., the continuity of the Cathaysian bridge and the smooth continuity of the Paleo-Tethyan southern shelf.

One would expect that the creation of the Cathaysian bridge might also have created a distinctive Paleo-Tethyan zoogeographic realm. The one problem here is that the bridge was almost continuously flooded in shifting corridors by shallow waters and thus enabled Paleo-Tethyan–Panthalassan exchanges at the littoral to shallow neritic bathymetric levels. Despite that, it has indeed long been pointed out that many genera and species in the Permian could be grouped into "Tethyan" biogeographic provinces (e.g., Ager, 1971, and Stehli, 1973, for brachiopods; Gobbett, 1973, for fusulinacea). In discussing the Tethyan provinces one must bear in mind, however, that this epithet is commonly (and regrettably) used as an equivalent of "equatorial," especially in the Permian world, and includes the lower latitudes in southern and western North America. Ross and Ross (1991, p. 361) noticed that "The latest Permian (Djulfian) saw a burst of diversity in the Tethyan realm and this produced some distinctive and successful lineages. These and the remaining survivors of the Guadalupian, suffered extensive extinctions before the end of the Permian;" but this statement again does not distinguish between the exotic tectonic units of the North American Cordillera and the Tethyan realm proper and thus helps us only in a limited way to circumscribe a true Tethyan realm zoogeographically.

However, more recent work in east Asia has shown that it is indeed possible to distinguish peculiarly Tethyan marine invertebrate faunas from those of the Panthalassa even at equatorial latitudes. Shi (2006), for example, was able to define four major provinces in eastern Asia on the basis of Permian marine faunas: Verkolyman, Sino-Mongolian-Japanese, Cathaysian, and Panthalassan. Of these, the Cathaysian realm is a part of what many paleontologists have so far called the Tethyan province and embraces the Paleo-Tethyan margin of the Cathaysian bridge (see fig. 3 in Shi, 2006), whereas the Panthalassan province includes all elements that flanked Pangea to the east (including the Panthalassan margin of the Cathaysian bridge!) and that had been transported from south to north along strike-slip faults within the Nipponide orogenic system during much of the Mesozoic (compare fig. 3, inset, in Shi, 2006, and fig. 21.53D in Şengör and Natal'in, 1996).[24]

Similarly, Angiolini et al. (2005) were able to distinguish a post-Asselian "Westralian" brachiopod province including western Australia, peninsular India, and the Himalaya (i.e., Tethyan) from an "Austrazean" province embracing only eastern Australia and associated Panthalassan units, i.e., forming a "Panthalassan" province in the southeastern part of the Permian Pangea.

There is, thus, little doubt that the late Permian world, in fact much of the Permian, had a Ptolemaic Earth, with the giant Panthalassa and the much smaller and essentially land-locked, equatorial Paleo-Tethys as two independent oceans. The next two chapters are devoted to documenting how anoxia developed in the enclosed ocean basins and their shelves. The first chapter discusses in a summary fashion anoxia and its products as a background and the second chapter reviews whether and where such products are seen in the Paleo-Tethys.

---

[23]The tetrapods of the Val Gardena Formation of the Southern Alps (Conti et al., 1977) and the report of a new find in northwestern Turkey that is currently in preparation by J. Sébastien Steyer, Sophie Sanchez, Okan Tüysüz, A.M.C. Şengör, Mehmet Sakınç, Olivier Monod, Erdal Kosun, Seveket Sen, Georges Gand, Ronan Allain, Fabien Knoll, Jean Broutin, and Martine Berthelin under the title "Permian tetrapod trackways from Eurasia show coastal migration-routes over Pangaea" may add some qualification to this statement.

[24]Incidentally, the presence of the Panthalassan zoogeographic province in the Permian also adds zoogeographical justification to the definition of the Nipponides.

CHAPTER V

Anoxia and its products

Anoxia is simply the condition of nonexistence of oxygen, prohibiting aerobic life to flourish. Etymologically, it means nonoxic; hypoxia (low oxygen), suboxia (under oxygenated), and/or dysoxia (badly oxygenated) are used commonly to denote degrees of oxygen deficiency above total anoxia. Berner (1981) classified sediments laid down in waters with greater than 30 μM of O_2 as oxic and those in waters less than 1 μM of O_2 as anoxic. Any sediment deposited in waters between these two values would be suboxic or dysoxic. He further subdivided anoxic sediments into those with less than 1 μM of H_2S as nonsulfidic and those with more than that as sulfidic, although major anoxic basins almost always have sulfur-producing bacteria (a remarkable exception is the Santa Barbara basin, offshore California, ~50 km × 100 km: Robert A. Berner, 2008, personal commun.)

Anoxia and dysoxia can happen both in the atmosphere and in various water bodies making up the hydrosphere, including lakes and oceans. Rivers, by their dynamism, cannot develop anoxia because they get mixed vigorously and whatever falls into them eventually gets carried away, except in abandoned ox-bow lakes within a wide flood plain. Anoxic condition affecting a substantial body of seawater was first discovered in 1890 during the Black Sea expedition of the Russian oceanographic vessel *Chernomorets,* and it was immediately recognized to result from inhibited internal circulation of the closed Black Sea basin (Andrusov, 1890); also see the preface in Yanko-Hombach et al. (2007, p. xi), for the current state of knowledge on the Black Sea redox conditions, see Murray et al. (2007). It was only after the discovery of black shales in the deep ocean during the Deep Sea Drilling Program (DSDP) Legs 36 in 1974, 40 in 1974–1975, 41 in 1975, 43 and 44 in 1975, 47 in 1979, 50 in 1976, and Legs 71–75 in 1980 that significant time intervals of ocean-wide anoxia in the past were recognized (Weissert, 1981; Okada and Kenyon-Smith, 2005, p. 164ff.).

Oxygenation depends on the extent and efficiency of photosynthesis and how its products are balanced by the multi-million-year carbon and sulfur cycles (what Bob Berner calls "long-term carbon cycle," i.e., how organic molecules are made, preserved, and destroyed: Berner, 1999, 2004, esp. p. 5–9, 2005; Berner et al., 2003, 2007). If there are oxygen-producing organisms (mostly phytoplankton in the sea and vascular plants in the atmosphere) around in excess of oxygen-consuming organisms and organic debris to be decomposed, i.e., to use oxygen, then the O_2 in the environment will increase. In the Phanerozoic, atmospheric oxygen levels hit an all-time high (±35%: see Fig. 18) following an episode of abundant vascular land plant reign in lush forests throughout a vast portion of the land surface in the early Carboniferous (Berner, 2003; Berner et al., 2007; Raymond et al., 1985; Rowley et al., 1985). The peak of oxygen abundance in the atmosphere was reached at a time of continental assembly and consequent aridification of the world climate leading to increased plant diversity but a decrease in absolute numbers of individuals in the late Carboniferous and the earliest Permian, ca. 280 Ma ago (Berner et al., 2007; for syntheses of the flora see Dimichele et al., 1985; Rowley et al., 1985; Willis and McElwain, 2002), when pests and diseases affecting arboreal life probably had not yet evolved to their subsequent levels. Also at this time, along the collisional systems of the Huastecan–Appalachian–Hercynian–Uralian, Altaid, and New England (in Australia) orogenic belts, rapid loss of long stretches of subduction zones diminished volcanic activity, contributing globally to a reduction in atmospheric CO_2 (see Beerling and Berner, 2005, for a systems analysis of the CO_2 fluctuations in the latter half of the Paleozoic, which is complementary to what is said here).

With the continuing formation of Pangea, aridity increased notably and forests were replaced by deserts and oxygen production thus diminished. The abundance of redbeds and coal of this age suggests consumption of large quantities of atmospheric oxygen (Berner, 2005), so much so that within 20 Ma its abundance plummeted to a low of 15%, with major evolutionary consequences (Berner et al., 2007). The steepest plunge took place between ca. 258 and 242 Ma ago (Fig. 18).

Changes in the atmosphere always generate their counterparts in the marine realm. First of all, when atmospheric oxy-

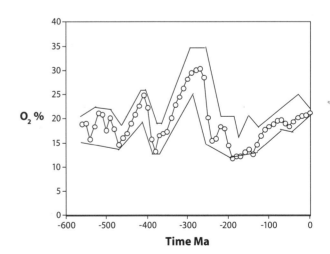

Figure 18. A curve showing the estimate of the amount of oxygen in the atmosphere according to Berner (2005). The curve shows the tremendous plunge the oxygen content of the atmosphere took in the Permian. We contend that this diminution must have exacerbated the effects the Paleo-Tethyan gas eruptions may have had on the terrestrial biosphere surrounding it.

gen diminishes, the contribution of the atmospheric oxygen to the aeration of the world ocean becomes reduced. Reduction of forests increases runoff in exorheic regions and thus desalinates the upper layer of the ocean, inhibiting circulation and thus aeration. In endorheic regions, by contrast, evaporites may form and get buried, further extracting salinating agents of the shallow ocean from surface circulation. In fact, some of the world's largest deposits of salt are of Permian age: Zechstein in central Europe and Kungurian in the pre-Caspian basin in southeastern Europe, Artinskian to Kazanian (±Roadian–Lower Wordian) in the Chu-Sarysu basin in Asia, and for much of the Permian in the Midcontinent basin of the United States (Zharkov, 1984). The presence of large salt pans would reduce iron transport by winds from continental surfaces to the ocean, thus helping to undercut the presence of oxidizing agents at the sea bottom and thus aiding the development of anoxia (e.g., Mahowald et al., 2009).

However, none of these factors are sufficient to create an anoxic ocean. In other words, a poorly oxygenated atmosphere cannot, by itself, generate an anoxic ocean or parts thereof. One needs the help of the biosphere and the lithosphere to produce efficient anoxia.

The lithosphere helps by influencing the physical geography; it helps to alter climates and changes the morphology of its own upper surface. By removing continents from the poles and allowing circum-equatorial currents to reign, the lithosphere deactivates ice ages, thereby increasing global temperatures. Oxygen solubility in water decreases with increasing temperature. For instance, a liter of seawater at 30 °C holds about half as much oxygen as at 0 °C (Vallaux, 1933, p. 729). Therefore, warmer climates result in less efficiently oxygenated waters. An end-

ing ice age also eliminates sources of cold, highly saline waters (Martinson and Pitman, 2007). Absence of these in turn inhibits global oceanic circulation. Thus, in greenhouse Earth times, less well–oxygenated waters cannot circulate as efficiently into the deep ocean as the better-oxygenated waters of the icehouse times. If, in a greenhouse world, the world ocean is divided into smaller closed basins, especially if partial or complete Ptolemaic conditions exist, the circulation in the closed basins will be even less efficient than the global oceanic circulation (Weissert, 1981). If, in addition, these basins are appropriately placed on the surface of the planet (e.g., in a hot region receiving many rivers) to generate very efficient thermoclines and haloclines, their total circulation may be stopped entirely, forming chemoclines as, for example, has happened in the present-day Black Sea (Fig. 19) and parts of the Mediterranean during the Messinian salinity crisis (Ryan and Cita, 1977; Cita, 2006; Ryan, 2009). Indeed on the present-day Earth, all anoxic areas are away from the mainstream of the global circulation (Fig. 20A and 20B).

What happens when anoxia is obtained? Basically aerobic life stops and anaerobic bacteria take over. Decomposition of proteins (animal or plant tissue) releases hydrogen sulfide gas (stink damp or sewer gas) following the reaction of carbonic acid with alkaline or alkali-earth sulfides. Hydrogen sulfide gets oxidized in the surface waters and dissolving calcium carbonate (calcite) gives rise to calcium sulfate (gypsum). The reduction of this calcium sulfate in the presence of organic carbon (dead organisms) generates calcium sulfide (calcium-oldhamite) and carbon dioxide, which, when it reacts with water, makes carbonic acid (soda water). The carbonic acid then reacts with calcium sulfide and generates calcium carbonate and hydrogen sulfide (Meunier,

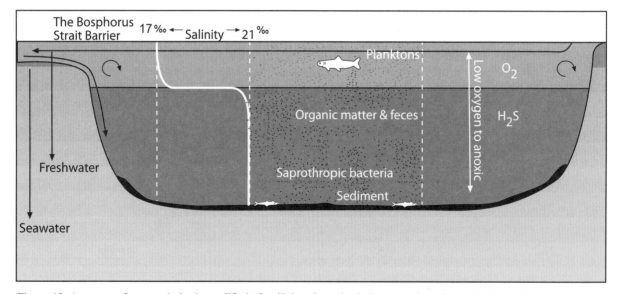

Figure 19. Anatomy of an anoxic basin modified after Holger Lueschen's interpretation of the Black Sea (http://eagle.icbm. uni-oldenburg.de/~mbgc/HolgerL/BlackSea.html seen on 4 January 2008). The influx of the heavy Mediterranean water from the Bosphorus settles on the bottom and inhibits circulation. The surface waters have just about half the salinity of the bottom waters and they cannot be mixed with them. H_2S and CH_4 accumulate at the bottom through organic debris accumulation.

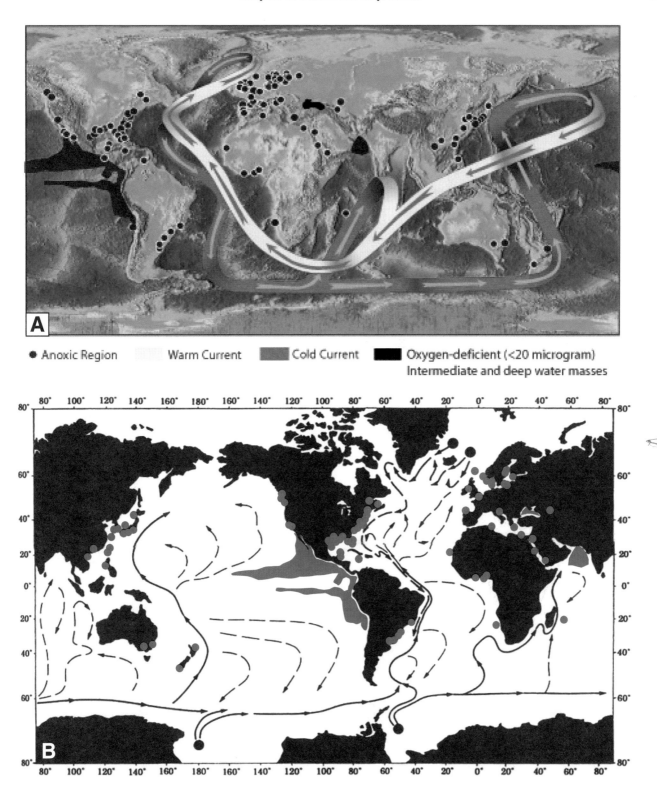

Figure 20. Regions of anoxia on the present-day Earth. (A) Places of anoxia with respect to the general circulation pattern in the world ocean (main current pattern after Gore, 2006, p. 151; oxygen-deficient water masses, i.e., the so-called "dead zones" after R.J. Diaz and R. Rosenberg, 2008, personal commun., and Potter et al., 1980, fig. 3.12). (B) Same as A, but with respect to the deep currents (deep current pattern from Stow and Lovell, 1979). The *Discover* Magazine March 2000 issue reported, in an interview with Robert J. Diaz of the Virginia Institute of Marine Science, that these "so-called 'dead zones' occur where fertilizer and sewage runoff, carried by rivers into the ocean, feeds the runaway growth of algae. When the tiny plants die, they decompose, drawing oxygen from the water. Fish can flee, but most sedentary creatures such as mussels, lobsters, and clams cannot survive." As we shall see below, this reflects precisely the pattern of the Permian extinction in the Paleo-Tethys.

36 A.M. Celâl Şengör and Saniye Atayman

1917; Erinç, 1984). If we bring more carbon and sulfur into the system than we can oxidize, the reactions will continuously produce carbon dioxide and hydrogen sulfide.

Methane is another gas that forms as a result of the degradation of organisms under anoxic conditions by iron, manganese, and sulfur reduction (Schubert et al., 2006). Formation of methane releases water or carbon dioxide through a complex series of chemical reactions. It has long been assumed that there may be three sources of methane in the water column in the oceans: contribution from sediments, contribution by methanogenesis in the water column itself, and contribution through seeps and mud volcanoes. The study by Schubert et al. (2006) has shown that methanogenesis in the water column in the anoxic Black Sea is negligible. The sediments at the bottom of the Black Sea are sinks rather than sources for methane in the water column, in contradiction to the claim by Reeburgh et al. (1991). Therefore, the major contribution must come from seeps and mud volcanoes, which the measurements of Schubert et al. (2006) corroborated. This contribution has been sufficient to raise the methane content within the anoxic part (i.e., below the chemocline) in the Black Sea to above 10 μM, which is, however, less than 1/150 of the necessary concentration to start exsolving (Yamamoto et al., 1976).

If all three gases (carbon dioxide, hydrogen sulfide, and methane) end up accumulating in very large quantities in the water column over millions of years, such conditions as increasing the temperature of the water or stirring it up vertically will lead to bubble formation and may eventually trigger gas eruptions as Kling (1987), Busson and Cornée (1996), Zhang (1996), Ryskin (2003), Kump et al. (2005), and Zhang and Kling (2006) argued. Ryskin (2003) and Kump et al. (2005) further pointed out that such ocean eruptions may have caused the end-Permian extinction, a thesis we believe very feasible.

Iron is also cycled through the oceans. If, in a marine environment, iron bonds more with sulfur, produced by anaerobic bacteria not common in well-aerated environments, than with chlorine, which is abundantly available in any marine environment, it is a good indicator that the environment is deficient in oxygen. The ratio of iron locked up in pyrite to pyrite iron plus reactive iron is called the degree of pyritization, and, if it is more than 0.75, the environment is considered anoxic. If the total iron in pyrite, in

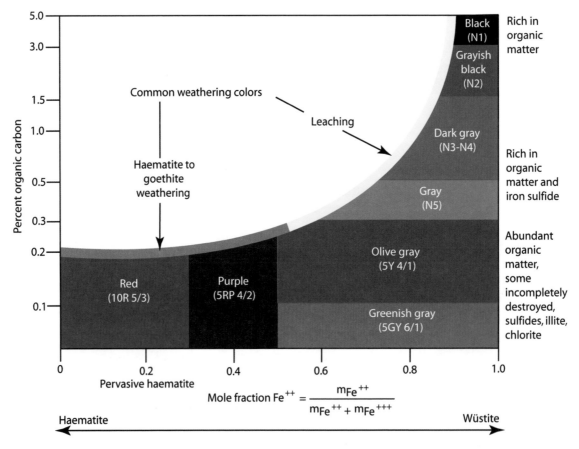

Figure 21. Suggested relationship of shale color to carbon content and oxidation state of iron, slightly modified from Potter et al. (1980, fig. 1.25) by addition of mineralogical data from McBride (1974). The mole fraction is used to indicate the proportion of the total iron that is in the +2 state and m represents the number of moles of iron per gram of rock. The Geological Society of America rock color chart (Goddard et al., 1975) was used for the color designations.

ankerite, in other carbonates, and oxides is more than about half of the available total iron, the system is also said to be anoxic because it could not use up all the iron to make oxides and carbonates. Geochemists consider that if the degree of anoxicity, defined as above, is more than 0.38, the system is considered anoxic. Slomp and van Capellan (2007, p. 162) define the degree of anoxicity in a different, but more dynamic way, as a function of vertical mixing rate in the oceans and net oceanic primary productivity, i.e., the rate of making organic compounds via photosynthesis:

$$\text{Degree of anoxicity} = 1 - k_{OA} \, [v_{mix}/F_{13}],$$

where k_{OA} is an apparent constant, which is proportional to the average dissolved oxygen concentration of downwelling surface waters, v_{mix} is vertical mixing rate, expressed as a dimensionless number by being normalized to the present-day mixing rate of 3 m/a, and F_{13} is net oceanic primary productivity. It is clear that in cases when downwelling surface waters carry little oxygen, vertical mixing is small, and primary productivity is high, there will be more anoxicity in the oceans.

In anoxic waters, the sediments will be generally replete with unoxidized organic carbon, sulfur, and iron. All of these impart a black to gray or grayish brown color onto the sediment. When rocks are black, they have much unoxidized organic carbon (up to 30%). If they are black to gray, then they have organic carbon plus authigenic iron sulfide (McBride, 1974). These conditions are generally interpreted to indicate anoxia (Wignall, 1994, and, following him, Strauss, 2006). In unmetamorphosed sedimentary rocks, green color indicates absence of both haematite and organic matter; it is imparted in part by illite, which is a proxy for enough oxygen to make this mineral. Green color we interpret as a dysoxic condition indicator. Even if the color is imparted by glauconite it still generally indicates reducing conditions. However, this criterion cannot be applied to metamorphic rocks bearing chlorite because it is that mineral which dominates the color in such rocks. In greenschist facies, one must search for indications of unoxidized carbon. Graphite and thus gray color

under such conditions is a helpful guide. Red and purple suggest the presence of haematite, both as grain coating and as interstitial mineral growth, hence indicating proper oxidizing conditions. Brown rocks are more difficult to evaluate. The color may be produced by faint or localized iron oxide coatings or by sulfur (McBride, 1974). Ferrous brown will thus be reddish, whereas sulfide brown will be yellowish-gray.

In addition to metamorphism, color may be altered by weathering. Light color can also result from weathering (e.g., many organic-rich rocks are black or brown in subsurface cores, but the same rock-types are light-colored in outcrop because the organic matter has been leached). Red haematite may weather to goethite so as to give the weathered surface of a red sedimentary rock a yellowish brown color. Primary color is probably best preserved in fine-grained mudrocks. That is why we tried to choose sections mostly with mudrocks as the main rock types in the sections we describe.

On the basis of such considerations, Potter et al. (1980, p. 55) devised a color guide to composition of shales, which we modified as our Figure 21 by adding mineralogical information from McBride (1974).

In the descriptions of the next chapter, black, dark gray, and gray sedimentary rocks will be considered to have been laid down in anoxic conditions; olive gray and greenish gray rocks in dysoxic conditions; and purple and red rocks in aerobic conditions. These observations are supported, wherever available observations allow, by reports of framboid pyrite (Strauss, 2006), Th/U values that are less than 2 (Adams and Weaver, 1958), identification of cerium anomaly (Chai et al., 1992; Kakuwa and Matsumoto, 2006), and an increase in $^{34}S/^{32}S$. We also paid attention to sedimentary structures (such as undisturbed laminae indicating lack of bioturbation) and lack of trace fossils (i.e., bioturbating agents) as possible indicators of environments hostile to oxygen-breathing organisms. The literature at our disposal is not rich, particularly with respect to trace fossil records, but we mention it here as a guide to future researchers who may find the main hypothesis put forward in this book worth testing.

CHAPTER VI

Some Permian sections in and around the Paleo-Tethys

In this chapter we describe 25 stratigraphic sections from throughout the world in which late Permian anoxic to dysoxic conditions are seen (Fig. 22A). In selecting them the condition was that the latest Permian and the earliest Triassic record be present in them all (except in section 21 which we reproduce to emphasize Yukio Isozaki's negative evidence) and that they collectively represent all bathymetric intervals from the abyssal plains to the shallow shelves in the world of the late Permian. In the Tethyan realm, with the singular exception of the eastern Mediterranean basin, the ocean basins are no longer extant and the ocean bottoms have been subducted. The eastern Mediterranean, however, is of little help to us, because so far no one has been able to observe intact Permian and Triassic sediments from its floor or from its margins. We therefore selected sections from subduction-accretion complexes representing, at least in part, sedimentary rocks scraped off the tops of subducted oceanic floors. In these structures, because of strong deformation, it is commonly impossible to separate the pelagic shales of the abyssal plains from the pelagic members of the distal trench turbidites. In fact, it is very rarely possible to see complete and/or continuous sections in them. We therefore had to make do simply with the rock packages as they appear to us in subduction-accretion complexes, as others have done before us elsewhere (e.g., Noble and Renne, 1990, in the Klamath Mountains; Isozaki, 1997, in the Chichibu and Cache Creek accretionary complexes; Takemura et al., 2002, in New Zealand; Wignall and Newton, 2003, in the Cache Creek accretionary complex). Such limitations reduce our choices severely and we are thus confined essentially to four areas where we are sure of sampling the rocks of a Paleo-Tethyan trench.

In the following sections we describe the appearance of the products of anoxia in temporal and/or spatial contexts of morphotectonic units of the oceans. The sections are numbered from 1 through 25 corresponding to the locality numbers in Figures 22A and 22B. The main authorities we follow for the sections are given in the beginning of each description so as to obviate the necessity of repetitive referencing.

Rocks of the Paleo-Tethys

Trench Fills and Abyssal (?) Deposits

1. Dizi Series, Svanetia, Georgia (mainly from Adamia et al., 1982; Adamia, 1984, and the personal observations of A.M.C. Şengör in the summer of 1984, during the 27th session of the International Geological Congress, Excursion 008 in the then Georgian Soviet Socialist Republic). The Dizi Series,

exposed in the cores of the Inguri and Tskhenistskali anticlines along the southern slope of the Greater Caucasus, is a highly deformed and very little metamorphosed clastic sequence representing the most important exposure of pre-Jurassic rocks in the southern slope of the Greater Caucasus. It is interpreted to have formed in a subduction-accretion complex: Şengör et al. (1988) interpreted it as Gondwana-Land bound, but a new study by Natal'in and Şengör (2005) showed that it could not be separated from the large subduction-accretion complexes forming a large part of the Scythides and must therefore have been Laurasia bound. *In both interpretations, however, it is an accretionary complex that formed within the Paleo-Tethys.*

The Dizi Series consists of a carbonate-terrigenous succession, including Middle Devonian tuffs and tuff breccias of andesitic composition with thicknesses of up to 200 m and ?Middle Carboniferous liparitic and dacitic tuffs. It spans an age range from the latest Silurian or early Devonian to the earliest Triassic. Local investigators divide the Dizi Series into three suites (formations in the Soviet stratigraphic nomenclature: Anonymous, 1979): the Devonian Kirar, Carboniferous to Permian Tskhenistskali, and the Triassic Gvadarachi Suites. Section 1 in Figure 22A shows only the latter two suites. All of these were highly deformed before black shales of Sinemurian age, dated by ammonites, were deposited atop them unconformably.

As far as the observations reach, the Tskhenistskali Suite is ~600–800 m thick and reportedly conformably overlain by the Triassic Gvadarachi Suite. Where Şengör was able to observe it in the summer of 1984, under the guidance of Professor Shota Adamia, the deformation was such that any reported stratigraphic relationship appeared to him suspect. However, at every outcrop, he was shown the fossil localities, so that the ages of the individual rock packages carrying the fossils seemed well established.

The Tskhenistskali Suite consists of olive-green, crenulated phyllites, silvery gray and brown sandstones, local volcanic layers and marble lenses, flinty slates, and siliceous deposits. The fossils mostly come from the marble lenses, but, also, albeit much more rarely, from the clastics containing the lenses, and range from the Tournaisian (*Siphonodella crenulata*) to the Wuchiapingian (*Waagenophyllum indicum*). The corals and the fusulinids (e.g., *Codonofusiella* sp.) come from the neritic limestone blocks in the clastic rocks and clearly represent an entirely different milieu of deposition from that of the matrix. The overlying Gvadarachi Suite consists of black shales, sandstones, silicic deposits, and again limestone lenses. These deposits are said to contain both *Cyclina* (!), which cannot be the Cenozoic mollusk, but must somehow be a member of the insect order of the same name of

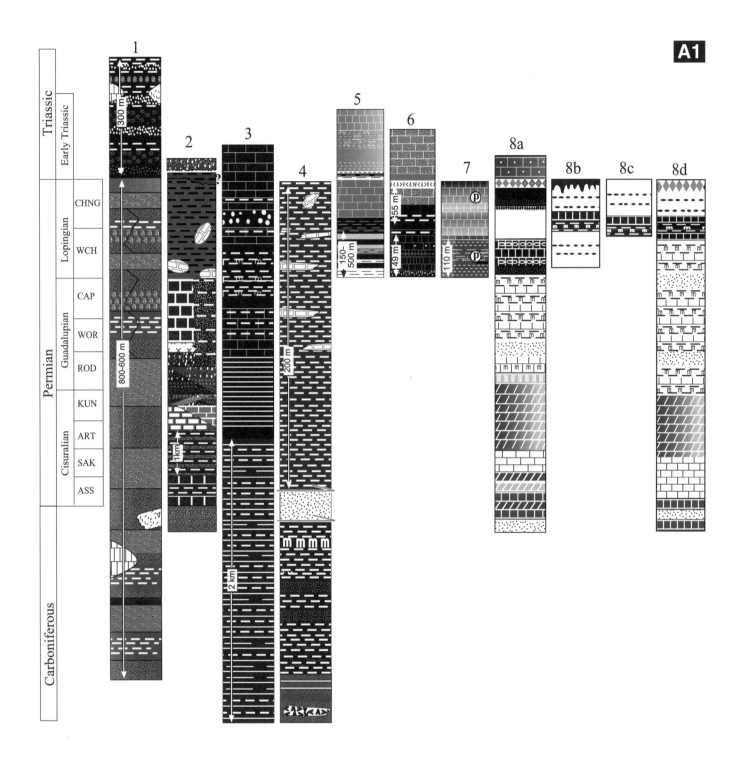

Figure 22. (A) Stratigraphic sections illustrating inferred redox conditions in and around the Paleo-Tethys during the Permian. Locations are shown in Figure 22B. The colors represent the actual colors of fresh rock surfaces as reported from outcrop observations. For sources and discussion, see text. ASS—Asselian; SAK—Sakmarian; ART—Artinskian; KUN—Kungurian; ROD—Roadian; WOR—Wordian; CAP—Capitanian; WCH—Wuchiapingian; CHNG—Changhsingian. (*Continued on following three pages.*)

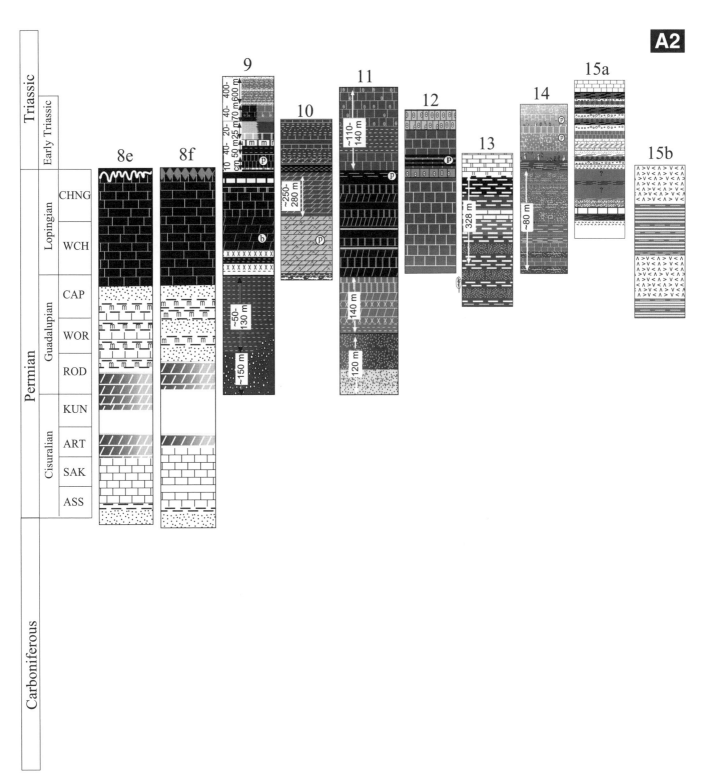

Figure 22 (*continued*).

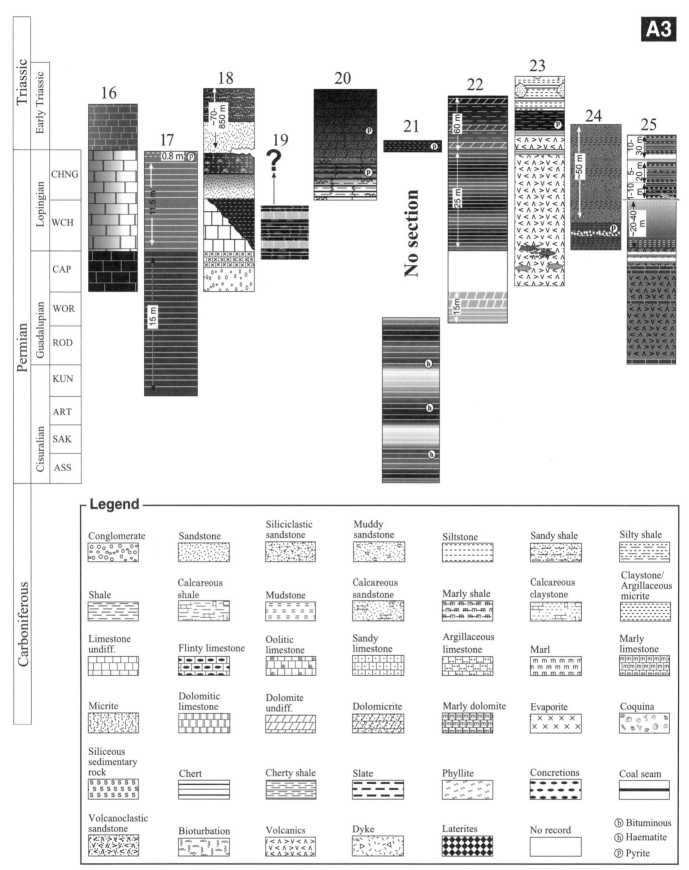

Figure 22 (*continued*).

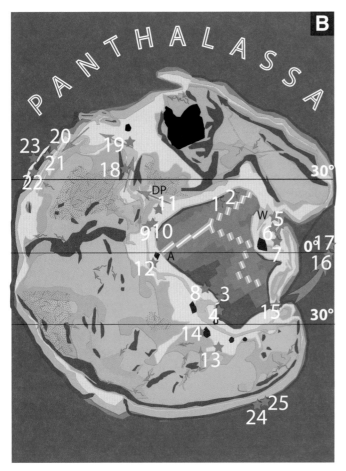

Figure 22 (*continued*). (B) Locations of the stratigraphic sections shown in Figure 22A in the late Permian world.

across a contact that is not exposed. The clastics pass into black limestones intercalated by argillite beds. Limestones and argillites collectively have thicknesses ranging from 10 to hundreds of meters. The age of these limestones range from the top Carboniferous to the Sakmarian. They are followed by about a kilometer of gray to dark gray, probably turbiditic shales and thinly bedded sandstones. These rocks appear azoic. Above them is, in places, a reefal limestone, whereas in other places thick, slaty sequences with dark color exist. The ages of the dark sequences of the "Beleulin-type" have been reported as Lower Carboniferous by Vinogradov (1958), but more recently Leven (1997) has seen fit to consider them Yahtashian (Artinskian to Lower Kungurian). In both cases, the clastic sequences are dated by fossils found in carbonates capping them, and the contacts in this severely disrupted environment may well be tectonic in both cases.

Above these clastics is the Gundara Formation shales and sandstones with tuff intercalations containing fusulinids and ammonoids. These rocks are variously dark colored and their observed thickness reaches nearly 2 km.

In the Murgabian (reaching into the Lower Guadalupian), sedimentation continued in some sections in a shallow water environment with red (secondary oxidation?) to multicolored tuffaceous clastic rocks, and, in other sections, in a deep water basin, where the younger sections of the "Beleulin-type" dark slates have been laid down. In the Capitanian, the entire area shallowed to the point where gypsum was being deposited in restricted basins and then rapidly redeepened to accumulate again dark terrigenous rocks containing plant, gastropod, and bivalve remains. Toward the central parts of the northern Pamirs, a Capitanian unit of black, bituminous, fusulinid-bearing limestone is underlain by conglomerates containing clasts from underlying reefal limestones.

In the northern Pamirs, it is not possible to distinguish the subdivisions of the Lopingian. It is represented by olive-green argillites containing limestone lenses, which we interpret as knockers in a shaly matrix and which contain Dzhulfian (≈Wuchiapingian) brachiopods. Therefore, the olive-green shales are at least of that age if not younger. The olive-green argillite grades into a red argillite and then is unconformably overlain by the Triassic conglomerates and sandstones. The red shales may well represent the Scythian as the overlying Triassic in Afghanistan begins with Anisian rocks.

In the northern Pamirs, the entire late Carboniferous to Triassic rock sequence dominantly consists of dark-colored clastics in

Asselian age,[25] *Catonia* sp., which is another insect genus, and *Dictyophyllum*, which is a fern genus that first appeared in the Carnian, known through its spores in the section. Although the milieu of deposition of the Gvadarachi clastic rocks was probably anoxic (black shales!), the fossils were clearly transported into it from nearby lands and shallow areas[26] and much mixed by the intense deformation. The entire Dizi Series seems laid down in anoxic conditions, with limestone blocks, which were deposited in entirely aerated conditions and clearly later incorporated into the anoxic sediments either as tectonic lenses or, in places, possibly as debris flow clasts, much like the famous *Calcari a Lucina* in the Miocene flysch deposits (such as *macigno* and *marnoso arenacea*) of the norther Apennines. We have before us a mélange, whose deep water rocks come from anoxic environments and its knockers mostly from very shallow water carbonate banks that probably sat on outer arc highs.

2. The northern Pamir Kwahan section (Leven, 1997, supplemented by Wolfart and Wittekindt, 1980). Here the section begins with brown marine clastics that sit on metamorphic basement

[25]This is an inexplicable situation because in the stratigraphic columnar section it is clearly stated that the Triassic Gvadarachi Suite contains Cyclina sp. (*sic*!) in Adamia (1984, p. 151). If it is an error in translation or transliteration, it is impossible to detect, because in the Russian version of the same book (p. 178), the same columnar section is given in Latin characters. The way we present it seems the only plausible interpretation, provided the authors did indeed write what we now have before us. It is also internally consistent, as insect fossils seem to occur where there are fern fossils in this area.

[26]It was first during the Challenger Expedition that it was realized to what great depths the debris of land plants and animals could be transported. In one case, the leaf of a shrub was dredged from a depth of more than 2000 m near the island of Palma in the Azores, and it was still green and firm! (Moseley, 1879 [1944], p. 506).

places capped by shallow water limestones that may or may not be black and bituminous. There is essentially no benthic fauna reported. Few pelagic fossils such as rare ammonoids have been collected from the dark-colored flysches. Where littoral facies are seen, the shales are red and in places even evaporites may be present. The impression one obtains is of deposition in a deep anoxic basin, in which rapidly evolving turbidite basins formed and deformed, and the deformed sections rose above the anoxic milieu to accumulate coral limestones. By Capitanian time, it seems as if anoxia had risen to depths in which limestone deposition had become possible. The picture is much the same as that we saw in the Greater Caucasus, nearly 3000 km away.

3. Central Pamirs (mainly from Norin, 1976, 1979). Here, in the central Pamirs and in the section extending also to the southeastern Pamirs across the Rushan-Pshart Zone, Ruzhentsev, whose observations Norin follows, divided the Carboniferous to the Triassic rocks into the following "suites": at the bottom is the Bazardar Suite ranging from Lower Carboniferous to the Lower Permian (Artinskian) and whose base is not exposed. The Bazardar Suite passes gradually into the Shindi Suite (Artinskian) and then into the Kubergand Suite, reaching from the Artinskian into the lowermost Middle Permian (Kubergandian). Then the Gana (Murgabian–Pamirian, i.e., from the Middle Permian to the Upper Permian), Karabeles (only Pamirian, i.e., going from top Gana to the top of the Permian), and finally the Kobrigen Suites follow, reaching into the Carnian.

Of these, the Bazardar consists of extremely monotonous, black and dark gray, laminated silty and fine sandy argillaceous distal turbidites and black shales. Their thickness may reach 2 km. Above them follow the rest of the suites up to the Triassic Kobrigen, and they consist of deep sea carbonates, marls, cherts, and shales all being black or dark gray, which Norin (1976) had an opportunity to examine himself through the cooperation of the Geological Institute of the Academy of Sciences of the USSR. The thicknesses of the pelagic sequences are very small: the Shindi, Gana, and the Karabeles together making up only 33 m in the Valley of the Boz Tere (Buff Creek, a name derived from buff-weathering dark shales?).

Here again, the entire section seems anoxic from the Lower Carboniferous to the top of the Permian, but, in this place, we are at the southern continental margin of the Paleo-Tethys, nearly 6000 km away from the northern Pamirs during the late Permian and south of the equator (Figs. 15B and 15C). The environment was deep marine, at least a continental apron, if not fully abyssal. The thinness of the deep sea carbonates, marls, cherts, and shales of the Shindi, Gana, and Karabeles Suites, to us, suggests an abyssal plain environment of deposition. The volcanics in the Shindi Suite are diabases and those in the Karabeles Suite are said to be volcanic ashes possibly spewed by a nearby magmatic arc. Given the state of disruption of the rocks, we do not hesitate to interpret the diabases as tectonic slices from the Paleo-Tethyan ocean floor.

4. The Horpa Tso section (Norin, 1946, 1976, 1979). The Horpa Tso (Turkic or Mongol Lake[27]) section was observed by Erik Norin in 1932, during the "Sino-Swedish Expedition to the North-Western Provinces of China" led by Sven Hedin (Norin, 1946). The section consists of two parts, to the north and to the south of the lake. Below we present a combined section:

The section begins at the base with half a kilometer of gray, greenish gray quartzitic arkoses intercalated by diabases and tuffs or devitrified lavas (hyaloclastite?). Above is a kilometer-thick clastic section that passes gradually from turbiditic sandstones to shales, with some chert layers, to a more distal turbidite sequence with quartz grains and small boulders in part carried by debris flows and some possibly dropped by icebergs, judging from their incongruent size and shape. Norin maintains that the quartz grains and small boulders were produced on a beach that experienced seasonal freezing. The turbidite consists of black to dark gray silt-shale intercalations with dark sandy slate horizons. Near the top, there is a marly limestone unit. Together, this represents Norin's Horpa Tso Formation, which he dates as Lower–Upper Carboniferous (Norin, 1976, 1979), although he repeatedly stresses the lack of fossils in the entire clastic section. Above is a tectonic contact bringing white quartzites. Across another tectonic contact are thick, grayish black turbidites with limestone knockers reaching a thickness of some 200 m, very similar to those seen in the southwestern Pamirs to the west. This is what Norin calls the Tashliq Kol (Stoney Lake) Formation. The section also has bluish gray crinoidal limestones with an age range from the mid-Permian to the Wuchiapingian, also probably of turbiditic origin, because it is quite impossible to imagine that the crinoids thrived in the depositional milieu of the rest of the deep-water section. Kahler (1974, p. 23), who identified *Praeparafusulina lutugini* Schellwein in this limestone, pointed out that most of these animals "certainly did not die of old age. ... During deposition, the tests were considerably damaged." In the section studied, the top of the section is crowned by a white, massive, dense, partly crystalline limestone rich in corals, bryozoans, and foraminifera with an age range from the Lower Permian to Murgabian (Upper Roadian to Wordian: see Norin, 1946). Blocks of bluish gray limestones in a calcareous matrix separate the white limestones across an area with no good outcrop from the dark turbidites and clearly mark a tectonic contact. The section near Horpa Tso greatly resembles that in the central and southwestern Pamirs (section 3) and, like it, represents a mélange formed in an accretionary complex that accreted the bottom sediments of the Paleo-Tethys. The clastic sedimentary rocks are similarly bereft of traces of life. The clasts in the Horpa Tso were being eroded from a cold region, in which there was seasonal sea ice, which, in part, explains the presence of arkoses in the lowest part of the observed section. With the exception of the entirely tectonically bounded white quartzite of unknown provenance and the Lower Permian to Murgabian white limestone knockers of shallow water origin, the entire Horpa Tso and Tashliq Kol sequence represents deposition in anoxic marine waters and corroborates the observations in the Pamirs.

---

[27]This is the highest freshwater lake in Tibet. It is also called Arport Tso or Gurmen Tso.

The conclusions about the Paleo-Tethyan trench fills and ?abyssal deposits are as follows: The four sections described above conclude what we could find concerning the abyssal areas of the Paleo-Tethys. Şengör has seen in the Longmen Shan in western Sichuan, near Wolong (see W in Fig. 15C for location in the late Permian world), late Permian dark, brecciated limestones topped by allegedly Triassic black shale and/or silt sequences forming the immensely thick turbidites of the Yajiang accretionary wedge (Fig. 15B, Ym). There is no reason why some of those clastics could not be top Permian (as far as Şengör was able to see they have no fossils). If so, it is worth noting that they are all black to dark gray. Similarly, the matrix of the mélanges from the southwestern margin of the Songpan-Ganzi System near Litang, in the westernmost parts of Sichuan, on the main road from Chengdu to Lhasa, indeed must contain deep water clastic rocks of Permian age. They, too, are almost entirely gray and black as Şengör saw them in 1993. In northern Turkey, the Aksu flysch (A in Fig. 15C) with an age range from ?Carboniferous to the Triassic (Şengör et al., 1984), interpreted to form part of a Paleo-Tethyan, Gondwana-Land-bound accretionary complex, is also dark gray to black. Here, both Güner (1979–1980) and Çağatay et al. (1979–1980) pointed out that in this classical massive sulfide locality, known and exploited since antiquity, the mineralization commonly appears at the contact of the abyssal tholeiitic basalts, and black shales plus some graywackes belonging to the Akgöl flysch (Şengör et al., 1984). In fact, Şengör has seen nowhere any deep-sea sedimentary rocks with proven or estimated ages of late Paleozoic in the supposedly Paleo-Tethyan accretionary complexes to have any color other than dark gray or just black. From this, we hypothesize that already in the Carboniferous much of the future Paleo-Tethyan deep-sea volume had turned anoxic. By Permian time, the observations are firmer: almost no piece of Paleo-Tethyan deep sea sediment of Permian age has any color other than black or gray.[28]

Indeed Gaetani et al. (1995) have also reported entirely anoxic conditions from the shales of the somewhat shallower Wirokhun Formation in the lower Chapursan Valley to the immediate east of the tributary Kundil in the northern Karakorum Mountains (colored 1:150,000 scale geological map in Zanchi and Gaetani, 1994).[29] The Wirokhun is of Upper Permian age and had formed atop southerly tilted blocks that had subsided to bathyal depths during the Midian (Capitanian), indicating sedimentation in a previously oxic environment that turned anoxic when the environment as a whole subsided into greater oceanic depths within the Banggong-Co Nu Jiang branch of the Paleo-Tethys. Gaetani et al. (1995) rightly correlated this with other anoxic sections in the eastern Tethys reported by Kajiwara et al. (1994). Farther north in the Karakorum, we enter the realm of the Wakhan (Hayden, 1915) (Misgar [Desio, 1963]) Slates that record anoxic conditions at least throughout the Permian (Gaetani et al., 1996; especially the informative syntheses in Gaetani, 1997a, 1997b).

In the following sections, we shall look at the shallower areas to see whether we can see any shallowing of the top of the anoxic zone on the Paleo-Tethyan shelves. In other words, we shall determine whether anoxia in the Paleo-Tethys continued expanding volumetrically.

Continental Margin Deposits

The following three sections are located on the margins of the Yangtze block (Y in Fig. 15B).

5. Meishan Permian–Triassic outcrops ~100 km northwest of Hangzhou, Zhejiang and Anhui Provinces (Dubertret, 1963; Sheng et al., 1984, 1985; Wang, 1985; Yin et al., 1986, 2001; Ding, 1986; Bureau of Geology and Mineral Resources of Anhui Province, 1987). The section is divided into the Longtan, Changxing, Yinkeng, Helohgshan, and Nan Ling Hu Formations.

The Longtan can be divided into three subdivisions. At the base are shales and intercalated, thinly bedded coals. These are followed by gray shales, gray-bluish shales, and fine-grained sandstones containing economically exploitable coal seams. At the top are yellow sandstones rich in brachiopod fossils and gray shales with thin coal seams. The formation thickness ranges from ~150–500 m. The Longtan Formation was laid down in a northwest-facing (present-day orientation) littoral region, called the "Littoral swampy and littoral clastic rock intercalated with carbonate rock facies province" by Sheng et al. (1985, p. 66). The coals, formed in the coastal swamps, include the following plants: *Psymophyllum* sp., *Rhipidopsis* sp., *Gigantopteris dictyphyooides,* and *Ullmania* sp.

The Longtan Formation passes conformably into the Changxing Formation of Changhsingian age and consists of two members. Below are siliceous black shales formed as distal turbidites. This lower member is ~90 m thick, but in places may be as thin as 35 m. It is overlain by proximal to medial, thinly bedded and/or banded, richly bituminous, gray and black micritic limestone turbidites. The whole formation was deposited in what Sheng et al. (1985, p. 71) called "Basin siliceous and clay rock facies" representing a low-energy, anoxic environment.

The Changxing Formation passes into the overlying Induan (Griesbachian) Yinkeng Formation that begins with shales and becomes progressively richer in limestones. The first 5 cm of the

---

[28]The only exception we know from the literature is the occurrence of a "reddish" Kungurian chert collected in north-east Khorasan in Iran in 1988 by the late regretted Anton Ruttner on the road to Garmab, 12 km to the east of the village of Sang-e Sefid and 35 km east of Fariman (35°46′ N, 59°49′ E) towards Safid Sagak (35°39′ N, 60°05′ E). This chert yielded badly preserved conodonts including *Mesogondolella shindyensis*, *Mesogondolella gujioensis*, *Pseudohindeodus nassichuki,* and *Hindeodus excavatus* indicating a late Kungurian (uppermost Lower Permian) age (Kozur and Mostler, 1991). Kozur and Mostler (1991, p. 105) also think *Pseudohindeodus catalanoi* may be present, thus making the age a little younger, i.e., basal Wordian, but they are not sure of the determination. Although the sample is not entirely red and the fossils are badly preserved, we tentatively take this observation as evidence for deep-sea aeration and interpret it as follows: Anoxia must have started in the deepest parts of the Paleo-Tethys, i.e., in the trenches. This piece of chert must have been deposited at an elevation much higher than trench floors, possibly close to a spreading center, but below the calcite compensation depth (CCD). This is the latest time we can now put our finger on for abyssal aeration in the Paleo-Tethys and possibly represents the last aerated pools on high-lying pieces of the ocean floor.
[29]Gaetani et al. (2004) provide a superb, beautifully illustrated introduction to the geology of this little-known region.

shales are white and the subsequent 11 cm is black and they gradually pass into 20-cm-thick micrites and then into marls with a similar thickness. Finally, a 45-cm-thick limestone tops the formation.

Isotopic analyses by Riccardi et al. (2006) of the carbonate-associated sulfate in the Permian-Triassic boundary rock samples provide a detailed record of several isotopic shifts in $\delta^{34}S_{CAS}$ approaching and crossing the boundary, ranging from +30 to −15‰ (Vienna Canyon Diablo Troilite [VCDT], i.e., the international sulfur standard), with repeated asynchronous fluctuations at two locations, one of which is Meishan. They interpret the patterns of isotopic shifts to indicate a shallow unstable chemocline overlying euxinic deep water, which periodically upwelled into the photic zone. These chemocline upward excursion events introduced sulfide to the photic zone, stimulating a bloom of phototrophic sulfur oxidizing bacteria. The same group also reported a negative shift in marine inorganic carbon-isotope composition ($\delta^{13}C_{carb}$) at the Permian-Triassic transition beds (Riccardi et al., 2007) consistent with the same interpretation.

The section we describe records a transgression. Although the turbidites in it are distal, in nearby sections of the same age there are proximal turbidites indicating that the depositional environments of the advancing transgression were strongly dependent on local bathymetric conditions. The whole environment was drowned during the entire late Permian, and finally the anoxic waters of the Paleo-Tethys invaded it and they seem to have stayed during the earliest Triassic. The presence of gray shales even during the Longtan-time (Wuchiapingian) may indicate that the anoxic waters reached the Meishan region before the region was completely invaded by the sea.

6. Outcrops between Hanzhong (Shaanxi) and Guangyuan (Sichuan) (Dubertret, 1963; Sheng et al., 1984, 1985; Wang, 1985; Yin et al., 1986; Şengör et al., 1988; Okay et al., 1989; Bureau of Geology and Mineral Resources of Sichuan Province, 1991; Isozaki et al., 2004): The section includes the following formations from base to top:

The Wujiaping Formation is formed dominantly from detritic limestones. At the base is an ~3-m-thick felsic tuff and tuffaceous sandstone, followed by thinly bedded bioclastic limestones. The dominant rock type of the formation is the 49-m-thick, dark gray, massive limestone containing abundant flint nodules and lenses. Above are 2-m-thick, gray-colored, brachiopod-rich limestones; 3-m-thick, fine-grained shaly limestones; 6-m-thick flinty limestones; and, finally, 2-m-thick black shaley limestones.

The Wujiaping Formation was deposited in what Sheng et al. (1985, p. 66) called a "restricted marine platform carbonate rock facies province," located in the northern part of the present-day Sichuan basin. This region was subjected to collisional orogeny during the latest Permian and the Triassic, and the north-facing Atlantic-type continental margin of the Yangtze block was entirely destroyed. Collision-related orogenic deformation marched far into the platform. That is why it has not been possible to find continuous sections representing the continental margin. The environment represented by the Wujiaping Formation was a

shallow, quiet, warm, and restricted basin with stagnant waters. Dominant rock types deposited in this basin were the gray or dark gray to black clastic limestones.

The Wujiaping Formation passes concordantly into the Dalong Formation of top Wuchiapingian (*Araxoceras-Konglingites* zone) and Changhsingian (*Pseudostephanites-Tapashanites*, *Pseudotirolites-Pleuronodoceras*, and *Rotodiscoceras* zones) ages, which consist of 28 m of mainly claystones and lesser limestones. Dalong is partly correlative with the Changxing and, like it, shows a transgressive character. At the base are 22-m-thick, bituminous black claystones, locally with 10–20-cm-thick marl interlayers. They are followed by a 4-m-thick, micritic gray limestone. The final 2 m are rhythmically bedded and contain felsic tuff horizons at eight different levels.

The Dalong was deposited in what Sheng et al. (1985, p. 71) called the "open marine platform carbonate rock facies and basin siliceous rock facies province." Here, during the early Changhsingian, a variety of animals and plants were thriving in an open and shallow marine milieu with clear waters. The transgression beginning in the late Changhsingian deepened the environment and the shallowing anoxia of the Paleo-Tethys invaded it. We have, at the top, tuffaceous-micaceous claystones, hyloclastic tuffs, and sandstones with benthic brachiopods indicating a regression during the latest Changhsingian.

The following, latest Changhsingian, olive-green, faintly laminated and almost entirely azoic marls mark the end of the Permian. In only the first 5 cm of these marls, *Pleuronodoceras tenuicostatum*, *Pentagonoceras* sp., *Huananoceras* sp., *Hypophiceras* sp., and the bivalve *Claraia wangi* were collected indicating the latest Changhsingian. However, a Triassic *Ophiceras* sp. was also found here leading Isozaki et al. (2004) to stress the importance of establishing the precise age of this fossil. As the following limestones contain the conodont *Hindeodus parvus*, it is clear that they are of Griesbachian age.

The Permian outcrops between Hanzhong and Guangyuan seem to record a transgression that commenced in the late Permian and lasted until the latest Permian. The conditions were not dissimilar to those that reigned in northwest Zhejiang, but shallowing commenced earlier in northern Sichuan and southern Shaanxi. This is consistent with their greater proximity to the advancing nappe fronts of the eastern Qin-Ling orogen and the associated flexural fore-bulge.

7. The Xiejiacago and Huayun sections, Sichuan (Dubertret, 1963; Sheng et al., 1984, 1985; Bureau of Geology and Mineral Resources of Sichuan Province, 1991). The Longtan and Changxing Formations are here as well. The Longtan here consists of 110 m of gray, gray-yellow, gray-black claystones, calcareous shales, and hard siliceous limestones. Elsewhere it contains a sandstone-shale sequence that may have been laid down by turbidity currents. It contains pyrite crystals near its base. The siliceous beds are in the middle parts of the formation and both below and above are some coal measures (not shown in section 6 because of scale). Şengör has seen these coals in the field in September 1986 and their association with deep water

sedimentary rocks led Prof. Kenneth J. Hsü, who led the field party, to indicate that they may be pelagic coals formed by burial of floating dead plant material and perhaps not therefore necessarily indicating a rapid shallowing of the environment. In fact, in the foreland fold-thrust belt of the Xiangganzhe suture belt in south China, the Longtan Formation directly follows the Giu Feng Formation consisting of chert beds.

The following Changxing Formation is conformable with the Longtan. Here also it is Changhsingian in age and consists of gray-white, gray-brown, and gray-black bedded and massive limestones. These limestones contain black chert nodules and thin black chert beds. In the lower and middle parts, shaly material is abundant, appearing in the form of black-gray shale beds and marls. Pyrite is ubiquitous. The benthic fossils seen appear to have been transported from shallower levels. Crasquin-Soleau and Kershaw (2005) showed that the earliest Triassic sedimentary rocks here were also anoxic and represented a continuation of the latest Permian conditions with "holdover" ostracods from the Paleozoic, proving that so long as "Permian" conditions lasted so did its expiring fauna.

In south China, we see the same general picture we shall note in other continental margins of the Paleo-Tethys: Anoxia seems to have reached the shallower areas in China much later than the abyssal areas of the Paleo-Tethys, exactly as it happened in other continental margin localities around the same ocean, to which we turn our attention next.

8. Northeast Iran, eastern Alborz Mountains, south of Gonbad-ı Qabus (Jenny, 1977; Stampfli, 1978; Jenny and Stampfli, 1978). This is an area located on the north-facing Pacific-type continental margin of Iran (Jenny, 1977, figs. IV1.1 and V.1.1; Stampfli, 1978, p. 102), in which shallow water sedimentary rocks have been laid down. As Jenny's and Stampfli's figures show, there was some minor vulcanism. In sections 8a and 8d, the lowest Permian Aleston is the lowest recorded formation. It commences with white quartz sandstones with a calcareous cement, not recorded in our sections 8a and 8d. These pass into gray to dark biodetritic limestones, then to gray marls, followed by a gray- to brown-colored dolomitic-calcareous section and finally micritic, yellowish dolomites. In sections 8e and 8f, the lowest recorded formation is Dorud, but in its sandy facies, which corresponds to Lower Assellian. Upwards, the sands pass into shales and finally into thinly bedded, oncolitic, coralline limestones with brachiopods. All of these rocks are white to yellowish in color with rare reddish intercalations both in the limestones and in the sandstones. The milieu was obviously well aerated. In section 8a, the Aleston Formation is followed directly by the Dorud carbonate rocks. In sections 8a, 8d, 8e, and 8f, the Kuh-e Sariambar (mountain of yellow store) carbonates follow the underlying formations so far described concordantly and are mostly yellow, light gray, and reddish in color. The age of these is Upper Sakmarian to Kubergandian (to Middle Roadian inclusive, i.e., to Lower Guadalupian). In all our sections, except 8b and 8c, the Gheslag Formation of dominantly red shales and some intercalated sandstones surmounts the Kuh-e Sariambar Formation. Near the top

of the Gheshlag, rise above sea-level is documented by coaly and lateritic horizons. However, in section 8f, toward the top, there is a deepening of the environment of deposition and *Bellerophon*-bearing black micrites alternate with black shales. This is the top of the Permian here, equivalent to the famous *Bellerophon* Formation of the Southern Alps. In sections 8a and 8d–f, the Ruteh alternates with Gheshlag. Similar to Gheshlag, its lower and middle portions were laid down in well aerated basins and produced biogenic limestones. Toward the top, everywhere black limestones, laid down in thin beds, predominate. Finally in sections 8a–d, the black cherty limestones, shales, and sandstones of the Nesen Formation appear above the Ruteh, but in sections 8a and 8d, the Gheshlag again surmounts the Nesen indicating complex interfingering relationships in a shallow depositional environment.

The overall picture of northeastern Iran is that of a continental shelf that had remained well aerated until the Upper Murgabian (Wordian) to Djulfian (Wuchiapingian), when the anoxic milieu began encroaching upon the shelf. However, because the lower parts of the Gheshlag were still deposited in an aerated environment and because the deeper water and entirely black Nesen does not extend farther to the south than section 8d (it is no longer present in sections 8e and 8f), it is clear that during the medial Permian the anoxic waters clearly had not yet reached the topographically higher parts where the earlier sections of the Gheshlag and Ruteh were laid down. By the late Permian almost the entire shelf had been conquered by the anoxic waters.

9. Paularo area, the Carnic Alps (Carulli et al., 1986; see also Bernoulli, 2007). The entire Southern and Austroalpine realm constituted, during the Permian, an east-facing broad continental shelf of the equatorial Pangea above an extensional Paleo-Tethyan subduction zone (e.g., Laubscher and Bernoulli, 1977, fig. 4; Massari, 1986, fig. 2). In the Southern Alps, this shelf was above sea level until the late Permian and was occupied by the Lombardian Verrucano ("*Verrucano Lombardo*") and Val Gardena fluvial conglomerates and sandstones except in deeply subsided basins farther east (in present day geographical coordinates) where Sakmarian to Artinskian black shales and quartz sandstones with local marginal limestone conglomerates and breccias were laid down, surrounded by coeval and Middle Permian reefs in high areas (Ramovš, 1986; Broglio Loriga and Cassinis, 1992). Clearly, such deeply subsided basins were in communication with Paleo-Tethyan deeper waters and admitted their anoxic contents.

The section begins in the Paularo area with the Val Gardena sandstone via a transgressive basal conglomerate with a red matrix of most likely Upper Permian age (Cassinis et al., 2002).[30] It passes upwards into a 50-m-thick brick-colored,

---

[30]Mauritsch and Becke (1984) reported evidence for the Illawara reversal of 265 Ma from the Val Gardena redbeds in the Paularo section. This would clearly date the section into the Capitanian, i.e., to the top of the Middle Permian. However, as Cassinis et al. (2002, p. 8) pointed out, the exact placement of the paleomagnetic samples within the sequence is doubtful, and the results contradict the indications of biostratigraphy. We here tentatively follow the age assignment as Upper Permian on the basis of the biostratigraphic evidence summarized in Cassinis et al. (2002).

grayish green siltstone. The Upper Permian *Bellerophon* Formation opens with lagoonal, white, gray, and pinkish saccharoidal gypsum, alternating with blackish pelitic sediments, representing shallow lagoonal anoxia. These are followed by *Rauhwacke*, blackish dolomite, white-grayish dolomitic and autoclastic breccia, and yellowish, friable dolomitic marls of some 250–280 m. The uppermost Permian consists of blackish bituminous limestones and dolomitic and marly limestones. There are interbeds of up to 20-cm-thick, blackish, laminated marls near the upper parts of this upper limestone sequence. The Permian-Triassic boundary is marked by a 10-cm-thick, oolitic limestone and black siltstones forming the famous Tesero Horizon, named by Bosellini (1964), of the Werfen Formation, which is overlain by marly micritic limestones and black siltstones of the 40–50-m-thick Mazzin Member. Above the Mazzin, the general anoxic conditions disappear.

In the Carnic Alps, too, we see the general invasion of the Paleo-Tethyan shelf by anoxic waters sometime during the later late Permian.

As one goes west from the Carnic Alps into the Dolomites and reaches the gorgeous Bletterbach Canyon—the Rio delle Foglie of the Italians[31]—one also arrives at one of the classical paleontological localities of the Alps (see especially Conti et al., 1977; Farabegoli et al., 2007), where, in the beautifully sculpted Val Gardena sandstones Ceoloni et al. (1988) and Conti et al. (2000) reported a most remarkable mixture of both top Permian and bottom Triassic vertebrate ichnofauna. The Permian ichnogenera are represented by *Ichniotherium* and *Hyloidichnus*, whereas the undoubted Triassic taxa are *Rhyncosauroides* and *Dicynodontipus*. This kind of mixture is also seen in the invertebrates of the faunally impoverished Tesero Horizon of Bosellini (1964) farther east (e.g., Crasquin et al., 2008; see especially Pasini's establishment that the Permian fossils in the Tesero Horizon are not reworked: Pasini, 1985). In the eastern and the Southern Alps, the Permian-Triassic transgression developed from east to west and the Permian-Triassic boundary cuts across rock-type boundaries and thus beds (Cassinis et al., 1993). Their fossils indicate that the eastern and the Southern Alps constitute a region of a number of survivors and the "boundary-crossing" tetrapod fossils of the Val Gardena Formation indeed plot outside the *independently drawn* "killing fields" of the Paleo-Tethys as discussed below (see Fig. 24). However, Visscher et al. (2001) pointed out that all recognized plant taxa in the Val Gardena Formation became extinct at or close to the Permian-Triassic boundary. They ascribe this to the "profound effect of the Permian-Triassic biotic crisis on gymnosperm diversity in the late Paleozoic Euramerian floral realm" (Visscher et al., 2001, p. 121). We are unable to follow this statement confidently in face of the diachroneity of the Val Gardena deposition across the Permian-Triassic boundary.

10. The Brsnina section, the Karawanken (Southern Alps), Slovenia (Buser et al., 1986; Dolenec et al., 1999; Dolenec, 2005).

This is an area that has a stratigraphy much like that in the Paularo area in the Carnic Alps and represents the same Pangean shelf. In the late Permian, in Brsnina, the Karavanke Formation transgressively covered the Val Gardena clastics and now consists of evaporitic-dolomitic sequences ~270 m thick. A 5-m basal terrigenous and/or dolomitic section is red in color and passes upwards into an 80 m gypsum and/or dolomite intercalation deposited in an evaporitic lagoon. Today these rocks are yellowish *Rauhwacke*, essentially a vuggy calcite rock, the vugs having been left behind by Recent dissolution of the gypsum. Within this rock is an interval of gray bituminous dolomite. Finally a 200-m-thick section of gray dolomite follows, as high as the Permo-Triassic boundary marked by a maximum 1-cm-thick clay layer with high uranium concentration. The gray dolomites above the *Rauhwacke* are full of framboid pyrite indicating deposition in anoxic conditions clearly heightened during the deposition of the so-called "boundary clay." The section passes continuously into Lower Triassic evaporites and clastics that are red in color.

Here, again, the picture is much the same as the one we saw in northwest Iran and in south China: a well-aerated shelf, locally interrupted by broad and very shallow, partly anoxic brine pools, that was invaded by anoxic waters during the second half of the late Permian. The sulfur isotopic studies indicate also that the late Permian waters were anoxic (e.g., Klaus and Pak, 1974). This sort of shallow sea-bottom is seen today, albeit in a much cooler climate and therefore with no evaporites, in the Archipelago Sea of the northern Baltic Sea, where there is a complex spatial distribution of anoxic, dysoxic, and oxic bottom waters depending on water depth and bottom topography, in waters less than 100 m deep (see Virtasalo et al., 2005, fig. 1).

11. Bükk Mountains, Hungary (Trunkó, 1996, p. 122–129; Less et al., 2005; Hips and Haas, 2006; Haas et al., 2007): The Bükk Mountains represent the northeasternmost outcrops of the equatorial Pangean shelf facing the Paleo-Tethys during the late Permian. This section was deposited in a low-lying part of the Pangean shelf, which, in the Southern Alps, became progressively deeper from west to east. That trend continued in what is today Hungary. We think it significant that this deepening is toward the Dobrudjan port (DP in Figs. 15A, 15B, 17A, and 17B), the only opening through which the Paleo-Tethys communicated with the boreal seas of the Panthalassa.

In the Bükk section, the Middle Permian is represented by the Szentlélek Formation, divided into a lower Farkasnyak Sandstone Member and an upper Garadnavölgy Evaporite Member. The Farkasnyak Member consists of well-bedded light gray, grayish green sandstone and red and lilac mottled sandstone with limestone nodules in the upper parts. The Farkasnyak Member passes upwards into the green claystones, dolomites, and gypsum anhydrates of the Garadnavölgy Evaporite Member. The environment of deposition of the Szentlélek Formation is one of a very shallow littoral and/or neritic area in front of a desert hinterland (see locality 11 in Fig. 22B). The Upper Permian is represented by the 250–280-m-thick Nagyvisnyó Limestone Formation consisting of medium-bedded, black limestones intercalated with

---

[31]Both designations meaning "the stream of leaves." "Leaves" refer to the thinly bedded rocks into which the gorge was sculpted.

black marls. Some black cherts derived from sponge spicules and pyrite occur in this formation and the whole aspect of the rocks points to anoxic depositional conditions, despite the rich fauna (Posenato et al., 2005). The organic matter was probably transported from nearby lands by turbidity currents as in the early Cretaceous north Atlantic while the depositing waters remained unoxygenated (Weissert, 1981). Above the Nagyvisnyó Formation is an almost azoic boundary clastic section, some 10 cm to a meter thick, including dark gray marls and sandstones dated as late Changhsingian on the basis of *Indivisia buekkensis* and the foraminifer *Earlandia*. This is the basal section of the Gerennavár Formation, the rest of which consists mainly of thinly bedded, stromatolitic limestones that are gray, dark gray, and light brown with yellow dolomitic lenses (see also Posenato et al., 2005, fig. 3). The base of the Gerennavár Formation was deposited in entirely anoxic conditions that rapidly evolved into well-oxygenated outer shelf highs and less well–oxygenated shelf basinal areas behind the highs. It still had its own sparse fauna, however, which was described by Posenato et al. (2005), which makes one wonder how continuously anoxic the waters were in which the carbon-rich sediments were obviously laid down. In any case, the anoxic waters arrived in the Bükk Mountains some time during the late Permian and fairly rapidly disappeared during the Lower Triassic.

The conclusions about the Paleo-Tethyan continental margin deposits are as follows: In contrast to the Paleo-Tethyan oceanic deposits, the continental margin waters turned anoxic during Cisuralian to Changshainan times depending on their height from the abyssal plains. Those that lay very low were invaded already in the early Permian as in the case of the deep water basins in the eastern parts of the Southern Alps. Such basins located in the continental crust are extremely rare, however. Others were invaded first during the Guadalupian to Wujiapingian interval. Where detailed information is available, as in northeastern Iran, it is clear that such areas were located on the distal, lowest, ends of the shelves. Finally, entire shelves turned anoxic only in the Changhsingian.

Rifts of the Future Neo-Tethys

The rifts of the future Neo-Tethys all began their activity within the Paleo-Tethyan southwestern Pacific-type continental margin, most likely as marginal basins or at least rifts exploiting an extensional continental margin.

12. The Alanya Nappes region, Taurus Mountains, southern Turkey (Altıner and Özgül, 2001; Groves et al., 2005). This region consists of a group of nappes, one group of which derived directly from the southern branch of the Neo-Tethys in Turkey (see Şengör and Yılmaz, 1981), which today forms the eastern Mediterranean and which is surrounded by what Altıner et al. (2000) call the "southern late Permian foraminiferal biofacies belt" of the wide southern continental margin of the Paleo-Tethys. A small embayment around the Gulf of Antalya within the present eastern Mediterranean began closing in the Cretaceous and

formed a stack of nappes called the Antalya Nappes (Lefèvre, 1967),[32] which, to the east of the Gulf of Antalya, are surmounted by a yet higher metamorphic nappe, the Alanya Nappe (Özgül, 1976; Okay and Özgül, 1984).[33] Groves et al. (2005) described the Demirtaş section from the Antalya Nappes appearing in an erosional window beneath the massive metamorphic *traîneau écraseur* of the Alanya Nappe.[34] Here the section begins above a sole thrust with bioclastic limestones in which the color becomes progressively darker upwards from gray to dark gray. However, just below the Permo-Triassic boundary, a thin unit of oolitic limestones are light gray to pinkish in color indicating aeration in a very shallow environment. The lower Triassic here is made up of laminar stromatolites, the first meter of which contains streaks of dark organic material, but the overall rock color is gray. The anoxia is only present in the top Permian except the topmost oolite section. In the Triassic section are pyritic strains, some cubic, some small and framboidal; however, they are intercalated with stromatolites requiring oxygenated conditions. We think most of the pyrites were probably deposited later and were not primary depositional features as no unified anoxic section is visible. Professor Demir Altıner thinks that temporally short anoxic episodes may be related to local upwelling (2007, personal commun.). In the Çürük Dağ section on the eastern side of the Gulf of Antalya, the same oolitic limestones have been found by Crasquin-Soleau et al. (2004) precisely at the same level, and the ostracod extinction at the very end of the Permian was shown to have been more abrupt and more intense than previously thought.

13. The Qubu section (Shen et al., 2001, 2004, 2006; rock color information supplemented from A.W. Bally, 1980, personal commun.). This section is located 30 km north of Everest within the Tethyan Himalaya. Recently it has been intensively studied as one of the spots preserving an uninterrupted passage from the Permian into the Triassic. When it was being laid down, the Neo-Tethys had just started rifting in this part of the Tethyan realm and, as such, the section at Qubu is still properly also a Paleo-Tethyan south margin section. It is subdivided into two formations: the Qubu below and the Qubuerga above.

The Qubu Formation is formed dominantly from dark brownish-gray quartzose sandstones interbedded with black shales and brownish-gray silty shales containing a *Glossopteris* flora (Hsü, 1976). Its age had first been thought Wuchiapingian (Hsü, 1976;

---

[32]The Antalya Nappes have been studied in their type locality by the late regretted Jean Marcoux since the middle 1960s (see Marcoux, 1987). All present work on them is based on Marcoux's meticulous and superb mapping and rich harvest of findings. Marcoux believed that the Antalya Nappes are essentially a part of the Lycian nappe edifice and ultimately derive from the north Turkish Neo-Tethyan sutures. We disagree with that interpretation for reasons set forth already by Şengör and Yılmaz in 1981. Since then much has been published for and against Marcoux's interpretation, but the majority opinion, shared by many of Marcoux's colleagues such as André Poisson and Olivier Monod and another Tauride master, Necdet Özgül, now seems to favor a southerly origin for the Antalya Nappes.

[33]For an easily accessible overview of the geology of the western Taurus, see Gutnic et al. (1979).

[34]The *traîneau écraseur* actually consists of three separate nappes so that one should properly speak of the "Alanya Nappes" (see Okay and Özgül, 1984), but their final emplacement over the window occurred in one piece.

Jin et al., 1997), but then revised to be Maokouan (Late Guadalupian: Hsu et al., 1990; Li, 1983) based on the marine fossils from the overlying Qubuerga Formation.

The Qubuerga follows the Qubu Formation conformably and consists of a sequence of siltstones interbedded with shales and bioclastic limestones containing abundant brachiopods, bryozoans, and corals. This formation is divided into three members. At the base is a 118-m-thick sequence composed of brown sandstone, muddy siltstone, and shale. According to its brachiopod fossils its age is Wuchiapingian. Next is a 120.5-m-thick sequence of gray silty shales grading upwards into bioclastic siltstones interbedded with bioclastic limestones. Finally, there is an 86.3-m-thick uniform dark gray shale with bivalves, plant stems, and unidentifiable ammonoids. The lower part of the upper member is composed of black carbonaceous siltstones and siliceous mudstones with radiolaria. Upwards, the rock type changes first to black siltstone and shale and then to varicolored shales with numerous ammonoids in the top 50 cm of the Qubuerga. There are no brachiopods in this member, but gastropods such as *Bellerophon*, *Retispira,* and *Naticopsis (Jedria)* and bivalves, dominated by *Atomodesma variabili*, are found abundantly in the lower part of the member. The topmost 50 cm of the Qubuerga Formation contains numerous, poorly preserved, thin-shelled ammonoids indicating a shift from benthic communities to nectonic taxadominated communities. This ammonoid-dominated assemblage continues into the overlying dolostone unit of the Tulong Formation of earliest Triassic age as indicated by the conodont *Hindeodus* cf. *parvus*. However, because the fossils are not right at the base, the very base of the Tulong may well have been deposited during the very latest Changhsingian.

In much of the Himalaya, all the way from west to east, the Upper Permian is dominantly composed of dark to black clastic rocks. Below it, the color is variable; there are gray to black sedimentary rocks such as the *Fenestella* shales of the medial Carboniferous and pink Blaini-type limestones (Gansser, 1964). Up into the Permian, the darker colors become gradually more predominant as, for example, in the Infra-Krols, where the pinkish Blaini-type limestones grade upwards into dark shales and slates and finally into jet-black carbonaceous shales and slates (e.g., Gansser, 1964, p. 89). The whole picture greatly resembles those in northwestern Iran and south China and gives the impression of a shelf area, where the deeper parts fell prey to anoxia earlier, whereas higher parts became invaded by anoxic waters later.

14. Guryul Ravine, eastern Himalaya, Kashmir, India (Razdan et al., 1985; Brookfield et al., 2003; Wignall et al., 2005; Shen et al., 2006; Algeo et al., 2007). This is a section that is farther "cratonward" of the Qubu section in the Himalaya. The late Permian is represented by the Zewan Formation, overlain by the Khunamuh Formation of uppermost Changhsingian-Lower Triassic age. The Zewan consists mainly of gray phyllites and some limestones and brownish gray sandstones. The ammonoids *Cyclolobus walkeri* and *Xenoaspis* sp. appear in the lower part of the upper half of the Zewan, thus dating it to the late Wuchiapingian–early Changhsingian. Here, the oxygen diminution is indicated by the darker sediment colors and the tendency to dwarfism in the brachiopods. The uppermost Zewan is essentially azoic and indicates the onset of the exterminations[35] already occurring in the late Permian here, contrary to the claim of Wignall et al. (2005). The Zewan is overlain by the dark calcareous shales and argillaceous limestones of the basal 16-m-thick unit of the Khunamuh Formation, called E. The basal part of E is called E1, consisting mainly of gray, silty, calcareous mudstones and lesser thin interbeds of fine-grained quartz siltstones and fossiliferous limestones with dwarf forms. These are the so-called *Claraia* beds and have no sign of bioturbation in them. Upwards, the color of the rock becomes darker and facies deeper with the appearance of *Otoceras*; these rocks are labeled E2. The Permo-Trassic boundary is determined to coincide with the E1-E2 boundary because of the appearance of the conodont *Hindeodus parvus* near the base of the unit E2. Pyrite is rare near the base of the unit E1, but becomes frequent about a meter above the base and remains so up to the base of the overlying unit E3, the *Ophiceras* beds. The anoxia had already begun to dissipate by the time the E3 unit was being deposited. The extermination commences, as already pointed out, in the top Zewan; it is most intense in the E1 unit and begins to wane in the E2 unit.

The conclusions about the rifts of the future Neo-Tethys are as follows: We opened a section under the heading "rifts of the future Neo-Tethys" only to emphasize that whatever is found in them during the Permo-Triassic interval actually belongs to the southern, broad continental margin of the Paleo-Tethys. The history of anoxia here is substantially the same as that on the Paleo-Tethyan shelves, which is to be expected, because all Neo-Tethyan oceans opened by disrupting the southern shelf of the Paleo-Tethys.

Paleo-Tethys–Panthalassa Transition Region

The one place where we have definite evidence that the Paleo-Tethys actually opens into the Panthalassa is in northern Thailand, along the Nan-Uttaradit–Sra-Kaeo–Bentong-Raub suture. Its relation to the Chiang Mai basin is completely unknown because the critical areas are under the sea. That is why and because of their great proximity to one another at the scale of our Figure 22B we combined the localities 15a and 15b into a single locality 15 in our Figure 22B. We describe the sections separately below.

15a. Lampang, north Thailand (Şengör, 1986). A continuous passage from the Permian into the Triassic is preserved in the so-called "Volcanic Group" of the Lampang area in northern Thailand. The Volcanic Group commences with tuffs and continues upwards with green and red shales, grainstones containing abundant fusulinids, and massive white limestone breccias. Above the limestones are green shales and finally rhyolites that

---

[35]Originally we had written extinctions here, but one of our reviewers, Professor Demir Altıner, admonished us to say simply "extermination" and not "extinction" in a section description, because extinction should be based on more that just one section. He is right and we bow to his judgment.

terminate the Volcanic Group. The siltstones in the conformably overlying Lampang Group are red in color.

In northern Thailand, crustal shortening related to the initial contact between Annamia and Sibumasu had already commenced in the early Guadalupian–medial Guadalupian boundary and the Nan-Uttaradit–Sra-Kaeo–Bentong-Raub suture had begun forming. Although the final collision was to be delayed until the later Triassic, the areas in the future collisional area were rapidly shallowing. The oldest Permian sedimentary rocks we know in the area, preserved only in the deeper marine parts, are anoxic to dysoxic and were most likely in communication with the rest of the Paleo-Tethys (see Figs. 15B, 15C, and 22B). Southwards, in the Bentong-Raub suture zone, the so-called "Schist Series" represents a group of deep-sea rocks with cherts and distal turbidites, although its age is unknown except that it lies somewhere between the Devonian and the Triassic. For whatever their worth, Şengör's observations made in December 1983 in the company of Ian Metcalfe between Kuala Lumpur and Bentong in Malaysia revealed just south of the Genting Highlands, on the main road from Kuala Lumpur to Bentong, the following:

Thinly bedded distal turbidites with an average of 5-cm-thick brown-red sandstone beds are intercalated with 5–15-cm-thick black and gray laminated shales. A few kilometers farther to the northeast are highly disrupted black and red bedded cherts with intercalated gray to black shales with some porphyries. All of these are parts of the accretionary complex, filling the Bentong-Raub suture here, and indicate the presence of original anoxic deep waters sometime in the late Paleozoic. Above these the turbiditic brown-buff colored Triassic Semantan and the terrestrial red-colored Tembeling Formations form unconformable covers.

15b. Medial-late Permian transition in the Chiang Mai area of northwest Thailand (Wonganan and Caridroit, 2006). For a long time the Nan-Uttaradit–Sra Kaeo–Bentong-Raub suture was thought the only suture of the Paleo-Tethys in Thailand. In 1994, Ricou made the felicitous suggestion that the Central Gneiss Belt and the ophiolites near Chiang Mai may be another, younger suture. In this suture belt, some 110 km to the northwest of Chiang Mai, in the Pai region (Mae Hong Son Province), Wonganan and Caridroit (2006) found blocks of Upper Maokuan (Capitanian) to Lower Wuchiapingian finely bedded shales, tuffs, and greenish-gray, gray, to darker colored cherts. Here, too, these blocks clearly indicate proximity to tuff spewing arcs and the presence of anoxic waters during the time when the cherts were being deposited. In the cherts and the shales only radiolaria have been found.

The conclusions about the Paleo-Tethys–Panthalassa transition region are as follows: This is the narrow strait in north Thailand, which was most likely two parallel, narrow straits, the Lampang and the Chiang Mai areas, connecting the Paleo-Tethys to Panthalassa. Through these straits, communication was probably established at the latest by late Maokuan, i.e., Capitanian time. This late establishment, with respect to the time of the onset of anoxia in the Paleo-Tethys, may well be because of the filling of the straits by flysch and thus the establishment of bathyal environments instead of truly abyssal milieus.

Panthalassan Sections

Areas Directly Polluted by the Paleo-Tethys

Claims of universal anoxia in the oceans during the late Permian so far have depended only on two groups of observations: the deep water sections from the Panthalassa-facing accretionary complexes in Japan and the accretionary complexes of the North American Cordillera. Of these two geographically widely separated regions, only the Japanese sections and the Redding section in the eastern Klamath Mountains of northern California have any legitimate claim to be truly Panthalassan, because the other published sections represent sedimentation areas behind arcs separating them from Panthalassan ocean floor. Different Japanese sections and the Klamath section, on the other hand, show different histories of aeration of the Panthalassa requiring an explanation. We have selected two sections to discuss the aeration history of the Panthalassa as recorded in southwestern Japan, one in the Klamath Mountains and two additional sections from New Zealand that show no indication of late Permian or early Triassic universal anoxia in the Panthalassa.

16. The Kamura section in the Chichibu Belt (Ota et al., 2000; Ota and Isozaki, 2006). The section begins with the black limestones of the Iwato Formation with abundant Guadalupian Tethyan fusulinids plus crinoids, large bivalves of Alatoconchidae (Isozaki, 2006), and dasycladacean algae. Although the limestones are black, their faunal content is rich and mainly benthic. Above the 19-m Iwato Formation is the white to light gray Mitai Formation of Wuchiapingian to Changhsingian that shows possible dysoxia only in its lowermost section. All of these are conformably overlain by dark to light gray Triassic micritic limestones. It is clear that some possible anoxia is seen in the Kamura section during the Capitanian, but not later. This is the exact opposite of what we see on the Tethyan shelves. To be able to explain this sharp discrepancy, we need to look at the geographical path the Kamura section followed until it became incorporated into the Chichibu accretionary complex during the medial Jurassic.

A look at the Permian Pangea shows that the Hida nucleus onto which the Kamura section was to be later accreted was to the south (Pangean coordinates) of the north China block (Şengör and Natal'in, 1996). It was situated right "in front" of the north Thailand strait(s), the only moderately deep water connection the Paleo-Tethys had with the Panthalassa. When Paleo-Tethyan anoxia rose above the sill depth of the bathyal Thailand strait(s), which must have happened sometime during the medial Permian as we know from the history of anoxia on the Paleo-Tethyan shelves, it must have flowed out onto the Panthalassan floor. The history of anoxia in the north Thailand region clearly shows that anoxic waters were crossing it during the Upper Maokuan (Capitanian). It seems that the Kamura section was overwhelmed by these waters at that time, but pulled itself out of their reach later. The return of the anoxic condition there during the early Triassic is not relevant here, because we know that the early Triassic ocean was not universally anoxic.

The discussion of the Kamura section brings us to the question of the ability of the Tethyan larger lenthic foraminifera to cross large oceans liberally. Alve and Goldstein (2003) have shown that although such foraminifera may be transported for large distances as propagules that are able to remain dormant for long stretches of time, larger benthic foraminifera do not in fact show panoceanic patterns of distribution. Bhaud (1998) had emphasized that the spreading potential of polychaete larvae does not predict adult distributions. Alve and Goldstein (2003) thought that if Bhaud's observation is applicable to foraminifera, it may be because of hydrodynamic (i.e., current, etc.) and environmental factors. We think that it is highly unlikely that Tethyan larger benthic foraminifera could have migrated to randomly distant atolls in the Panthalassa; their presence on exotic tectonic units must therefore imply that these units were physically close to the Paleo-Tethys when the migration occurred. If true, this interpretation provides additional support to our view that the Kamura section was never very far from the Paleo-Tethys and may in fact have been located in front of the bathyal Thailand strait(s), where Paleo-Tethyan waters, together with their faunal load, probably engulfed it. That is why we are not nearly as sanguine as our friend Yukio Isozaki that the Kamura rocks actually are representative of the entire Panthalassan conditions (see Musashi et al., 2001; Isozaki et al., 2007).

17. Sasayama section (Kato et al., 2002). The interpretation given above of the Kamura locality is corroborated by the entirely divergent history of anoxia seen in the Sasayama section, pieced together from the same tectonic belt. In that section (Kato et al., 2002), there is sporadic anoxia until the Wuchiapingian, showing different spells of anoxic waters coming into the area of deposition (most likely gorging out of the north Thailand strait[s]). From Wuchiapingian onwards, we see a dominance of gray cherts and these become darker upwards. Here, clearly, the area of deposition was gradually overwhelmed by the anoxic waters in complete contrast to the Kamura area, possibly by different spells of outgorging of anoxic waters, perhaps similar to the episodic spells of disgorging of heavy Mediterranean waters into the Black Sea through the Bosphorus, as opposed to the continuous flow of the lighter Black sea waters in the opposite direction (Scholten, 1974). This picture is consistent with the interpretation that the whole of the Panthalassa had not turned anoxic and anoxia only affected those areas that happened to be near the Panthalassan mouth of the Thailand strait.

The conclusions about areas directly polluted by the Paleo-Tethys are as follows: Two areas located probably close to one another in the Pangean geography and also to the mouth of the Thailand strait(s) show entirely different histories of anoxic and dysoxic sediment deposition, which is most conveniently interpreted as repeated anoxic water disgorge through the Thailand strait(s) toward the Panthalassa or perhaps different times of passage of the two sections in front of the Thailand strait(s), if continuous flow of toxic waters are assumed through it.

Areas Indirectly Polluted by the Paleo-Tethys

These are the areas that have been fed by the Paleo-Tethys only through the complex rift chains across southeastern, central, and western Europe during the Permian (Figs. 17A and 17B). It is likely that the anoxic waters coming from the Paleo-Tethys were not nearly sufficient to pollute the entire water body in these places. They seem only to have helped to increase the anoxic tendencies of the low-lying basins filled with heavy brines there.

18. Jameson Land–Hold With Hope section, east Greenland (Bjerager et al., 2006). This superbly developed Permo-Triassic section has long been known to exist in east Greenland (Trümpy, 1960), superior indeed to the much condensed Meishan section in south China (Bjerager et al., 2006). The section commences with a continental, alluvial conglomerate that grades into marine sediments at its top. The marine sediments develop into the evaporites and the hypersaline carbonates of the Karstryggen Formation. Upwards, in the Wuchiapingian, the stratigraphy is differentiated into the shallower-water, ~150-m-thick Wegener Halvø carbonates and the black, bituminous shales with *Posidonia permica* Newell forming the 100-m-thick Ravnefjeld Formation. In the Changhsingian, the Schuchert-Dal Formation is the youngest Permian basin fill, being represented by gray silty bioturbated mudstones (Oksedal Member) and turbiditic sandstones (Bredehorn Member). During the time of the transition from the Permian to the Triassic, rifting continued here and those areas located on rotating foot walls rose above sea-level while others on sinking hanging walls continued to be submarine, and it is on those sinking blocks that sedimentation across the Permian-Triassic boundary remained continuous, as for example, in the Oksedal and the Aggersborg areas in northern Jameson Land. In these places, the Wordie Creek Formation overlying the Schuchert Dal Formation is still of Changhsingian age, albeit very latest Changhsingian (*Hypophiceras triviale* zone). This is the very area where there has long been a discussion whether Permian faunas had continued living in the Triassic (cf. Trümpy, 1960; Grasmück and Trümpy, 1969; for a discussion of this controversy, see Hallam and Wignall, 1997, p. 117). By contrast, Bjerager et al. (2006) document that the faunal turnover had taken place already in the latest Permian and that when the Triassic opened some diversification had already taken place. This is indeed an area of differentiated response of the marine biosphere to the events going on in the lithosphere. When the lower basinal areas were overwhelmed with heavier anoxic waters, it is there that extinctions took place. In the higher areas, on the upthrown fault blocks, aerated areas possibly harbored the survivors, but the sediments there were mostly eroded away. Where erosion did not destroy the evidence, one sees that the boreal Permian faunas continued living happily into the Triassic as Grasmück and Trümpy (1969) pointed out and as Professor Rudolf Trümpy still maintains (2007, personal commun.). As the old "survivors-argument" should long have alerted us, and as Bjerager has very nicely shown, a strict synchroneity of the faunal turnover in the boreal realm and in the Tethyan realm does not exist. In fact, the biospheric events in east Greenland seem intimately tied up

with the lithospheric and/or hydrospheric evolution of the east Greeenland Norwegian rift system.

19. Kap Starostin Formation sections, Spitsbergen (Małkowski, 1982; Nakamura et al., 1987; Mangerud and Konieczny, 1993; Wignall et al., 1998). A very similar picture to that encountered in eastern Greenland is present in Spitsbergen, from where Małkowski (1982) has provided very detailed litho- and biostratigraphic data later supplemented by further paleontology and geochemistry. None of these has been able to resolve the precise age of the Kap Starostin Formation, and estimates for the age of its top range from the earliest Tatarian (*Kraeuselisporiles* assemblage: ± Capitanian) to Dzhulfian (Wuchiapingian, on the basis of analogy with *Cyclolobus* beds of the Foldvik Creek Formation in east Greenland) on paleontological grounds. Wignall et al. (1998) claim a top Permian age for the top of the Kap Starostin Formation on the basis of the presence of the fungal spore *Tympanicysta stoschiana* in the basal part of the overlying Vardebukta Formation, but this is a circular argument that accepts a dramatic and singular mass dying only at the very end of the Permian, which has long been shown not to be true (e.g., Magaritz et al., 1988; Stanley and Yang, 1994). We here accept a Capitanian to Wuchiapingian age for the top of the Kap Starostin Formation (because the only tangible paleontological evidence points to those dates) and argue that its most complete sections in Kapp Starostin Festningen, Ahlstarndodden, Reinodden, and Polakfjellet localities simply indicate stagnant deep basins in which mostly black and gray cherts have been laid down and well-aerated shallower highs on which yellowish to cream-colored carbonates accumulated. This submarine topography was fault-controlled as Małkowski (1982, fig. 5) illustrated and indicates sporadic arrival, and further development because of the restricted basinal areas, of anoxic waters. There never was a "general anoxia" in Spitzbergen as there never was a general anoxia in east Greenland.

The conclusions about areas indirectly polluted by the Paleo-Tethys are as follows: This is a complex environment dominated by active rifting during the Permo-Triassic interval. We have seen above that the Carapelit rift in the North Dobrudja provided an anoxic water passage to interior Europe at most sporadically. We know also that some Tethyan animals passed through the Carapelit rift to appear in Silesia (Peryt and Peryt, 1977) and farther north (Merla, 1930). Northern Germany and the North Sea area were at the time a broad area of rifting and post-rifting bovine-head basin development. It is in these basins that the famous Zechstein evaporites were laid down. The Zechstein is now commonly divided into seven sequences in northern Europe, consisting from oldest to youngest of the Werra, Strassfurt, Leine, Aller, Ohre, Friesland, and Mölln (Henningsen and Katzung, 2002, p. 97), which have developed from the earlier fourfold subdivision generated after World War II (Richter-Bernburg, 1955). Of these, a sequence of dolomites, called the *Hauptdolomit* (main dolomite),[36] has been deposited as part of the Strassfurt sequence under anoxic conditions and led, in the potassium mine Volken-

roda-Pöthen, to a spectacular petroleum "eruption" in 1930 as a consequence of salt-mining-related decompression (Henningsen and Katzung, 2002, p. 115). Elsewhere, other sequences show evidence of anoxia and, for example, marcasite growth (Wahnschaffe and Schucht, 1921, p. 13). Regional variability is great in these shallow evaporitic pans, but in most places a sandstone-shale-carbonate-evaporite sequence is seen (e.g., Sokołowski, 1970, p. 530ff; Ziegler, 1990), determined by sedimentation along locally shifting platform-slope-basin configurations caused by an interplay of rifting and sedimentation (see, for example, the excellent paper by Geluk, 2000). The Werra sequence begins by desert sand dunes and rapidly evolves into the copper-shale that is rich in metal sulfides. Evaporites follow these. The Strassfurt sequence begins with the *Stinkschiefer* (stinking shale), a shaly dolomitic marl, or, in other places, by the *Hauptdolomit*, which is commonly gray or brown and bituminous containing locally exploitable petroleum. The higher sequences have dark clays at the bottom and become progressively more reddish upwards.

Figure 17A illustrates the "rift path" of the Paleo-Tethyan waters toward the Boreal realm. Figure 17B is a very schematic cross section along this path. During the pre-late Permian times there is no evidence for universal anoxia on the Tethyan shelves but areas such as the rift basins in the eastern Southern Alps were already invaded in the early Permian (see Fig. 23, below section 10). We know that the Carapelit rift began allowing Paleo-Tethyan waters to pass into the central European rift basins at the latest by Kungurian time. The Zechstein, however, is later Wuchiapingian in age (Menning et al., 2006), and it is at this time that we see widespread anoxia in the central European basins (well after the earlier *Kupferschiefer* anoxia). This may have corresponded to a situation as depicted in Figure 17B-b, i.e., to a time interval when the Paleo-Tethyan anoxia began invading the lower lying areas of the Paleo-Tethyan shelves. However, as the cyclicity of the sedimentation in the central European basins show, the water influx was not steady, but came in pulses, possibly influenced by episodes of rifting both in the future north Atlantic and along the port of North Dobrudja (Figs. 15B, 15C, 17A, and 17B), formed mainly by the Carapelit rifting. As Richter-Bernburg (1960, p. 134, footnote 1) already emphasized, in the Zechstein rhythmic sequences that begin with deeper water clastic and carbonate rocks and end with halite deposition, the sedimentation always begins in an anoxic environment (as the new waters come into the basin) and terminate in shallow, well-oxidized salt pans. This is compatible with episodic anoxic water arrival from the Paleo-Tethys into northern central Europe.

However, the stagnant, saline waters in central Europe must have encouraged further anoxia. As these waters were then spilled over into the future north Atlantic rifts, they would accumulate at the bottoms of the rifts and possibly combine there with the local, already developed anoxia in these restricted basins resembling giant fjords cutting into the corpus of the Pangea.

We think that water spilling into the central and northern European basins from the Panthalassa and from the Paleo-Tethys must have alternated in time, which means that as the

---

[36]Not to be confused with the Alpine *Hauptdolomit*, which is Triassic.

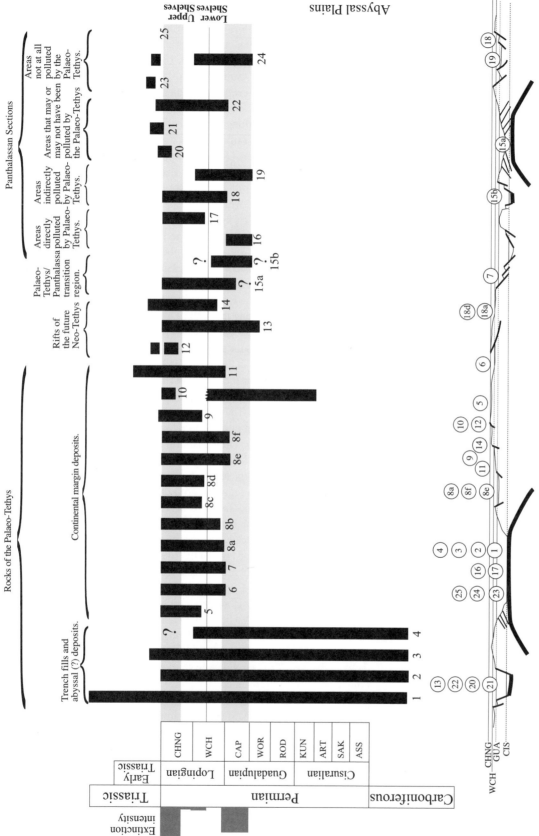

Figure 23. Summary diagram of the development of anoxia (black) in the sections illustrated in Figure 22A and its correlation with the Permian extinctions. In these sections, anoxia does not everywhere set on abruptly, but evolves gradually with initial dysoxic onset. We did not distinguish dysoxia leading to anoxia, but painted the entire section black. However, most of what we colored black consists of anoxic sedimentary rocks. The blue histogram to the left was taken from Stanley and Young's 1994 paper, figure 5, which shows the ratio of last to first appearances of brachiopods for the upper half of the Permian in China. It turns out that this diagram gives a fairly faithful picture of the entire Permian extinction pattern on land and at sea with insignificant variations. For instance, Spencer G. Lucas (2007, personal commun.) finds that the tetrapods experienced a major extinction by the disappearance of the Dinocephalians at the end of the Wordian or in the beginning of the Capitanian. This is not sufficiently different from the totality of the marine extinctions because the accuracy of biostratigraphy can hardly allow to distinguish the onset and termination of either. The bryozoans also had a protracted extinction that started in the Capitanian as Powers and Bottjer (2007) recently have shown. We feel that with accumulation of more data, the Permian extinction peaks will lose their sharpness and reflect more faithfully the progress of anoxia. The figure at the bottom shows a variety of tectonic environments (with no regard to any actual geography) and the times at which they became invaded by anoxic waters. The numbers in circles correspond to those of the sections above to show which section was deposited in what kind of a morphotectonic environment.

Paleo-Tethys thereby spilled into the central European basins, they too may have spilled into the Paleo-Tethys, bringing their saline waters into it and contributing to its further stratification. By latest Permian time, more anoxic waters must have been pouring into central and northwestern Europe as indicated in Figure 17B and 17C.

Areas That May or May Not Have Been Indirectly Polluted by the Paleo-Tethys

In certain restricted basinal areas in the North American Cordillera, anoxia developed in the late Paleozoic and the early Triassic (and also at other times) and some of these may have been fed by the anoxic waters coming from the boreal realm. However, this is hard to document. We mention them only to formulate a hypothesis that may be tested paleontologically.

20. Ursula Creek section, Lake Williston, British Columbia (Wignall and Newton, 2003). This section is located within the Cache Creek suture (commonly and, inappropriately, called the "Cache Creek terrane") of the Canadian Cordillera in British Columbia. The descriptions we used commence with the Fantasque Formation consisting of decimetric-bedded, gray cherts (http://www.empr.gov.bc.ca/, seen on 26 December 2007) and interleaved dark shales with centimetric layers. Using the presence of a negative $\delta^{13}C$ spike, the Permian-Triassic boundary was fixed here as being between the Fantasque Formation and the overlying Greyling Formation. The age of the Fantasque is dated as Changhsingian, and anoxia first fully appears toward its upper horizons and continues in full bloom into the Lower Triassic. The picture is in substantial agreement with that provided by Isozaki's (1997) descriptions farther south and shows the onset of anoxia much later than the Paleo-Tethyan sections.

21. The Cache Creek area (Isozaki, 1997). We deliberately mention this section from Isozaki, who used it to claim the existence of universal anoxia in the world ocean during the Permo-Triassic transition. The one remarkable thing about it is that there are *no* rocks preserved here that represent anything in the Lopingian or in the Lower Triassic. Isozaki (1997) claims that the rest of the section is similar to a constructed section showing anoxia in southwest Japan and therefore the missing bits should also be similar. This is an argument from *negative* evidence and inadmissible at face value. Isozaki (1997) uses only the evidence from the rest of the Permian-Triassic time interval to argue for universal anoxia. In the Cache Creek, ribbon cherts from the Permian to the Triassic are of tan, black, white, red, and green varieties as, for example, in the Atlin area (Mihalynuk, 1997, p. 35). Therefore, the *positive* evidence indicates a variety of environments of different degrees of oxygenation in the oceanic basin now represented in the Cache Creek suture zone during the Permo-Triassic and gives no indication of worldwide "superanoxia."

22. The Quinn River section, northwest Nevada (Sperling and Ingle, 2006). This is a part of the accreted Paleozoic arc complexes of Sonoman age in the western United States (Dickinson, 2004, fig. 6) exposed in the Bill Creek Mountains, some 97 km north of Winnemucca. The lithostratigraphic units are designated by numbers by Sperling and Ingle (2006) and their description begins at the base with a 2-m-thick bioclastic calcirudites overlying 15 m of cherty tuff, clearly indicating proximity to an active island arc, possibly already then facing North America. Above it is a 15-m-thick light brown dolomite containing Wordian brachiopods. It is nonsequentially overlain by 25 m of gray to black spiculitic radiolarian cherts. Above it are conformable 60-m-thick deep water Lower Triassic dolomites and shales and then black shales and cherts. The gray and black deep-water deposition began during the Wuchiapingian and continued upward into the early Triassic. Sperling and Ingle (2006) note the disappearance of the cherts in the late Permian and their return in the Spathian and correlate it with the alleged "global chert gap." The basin in which the Quinn River sequences were laid down were *not* Panthalassan basins, however, but more akin to the marginal basins of the present day western Pacific. Therefore it does not reflect the conditions reigned at the bottom of the immense Panthalassa. The so-called "chert gap" is only a gap because more than half of the surface of the planet, and especially that surface most suitable for the formation and preservation of cherts during the Permo-Triassic transition, is no longer available for observation. In fact, when Takemura et al. (2002) looked at the Permo-Triassic deep-sea deposits in New Zealand, they found no chert gap at all! As De Wever et al. (2006) rightly point out, there probably was a true decline of the plankton population during the late Permian, but there also is the dearth of stratigraphic preservation of the deep sea record.

Areas Not at All Polluted by the Paleo-Tethys

23. The Redding section, Klamath Mountains, northern California (Coogan, 1960; Noble and Renne, 1990). The Redding section—representing the record of an ensimatic arc growing outboard of North America—is one that is probably truly Panthalassan as there were no farther arcs outboard of it (Burchfiel et al., 1992a; Dickinson, 1992, see especially his fig. 3; Miller et al., 1992). The section contains the Dekkas Formation, an ~2-km thick pile of volcanics, volcaniclastics, and sedimentary intercalations consisting of varicolored cherts and shales. The petrography of the red Dekkas cherts revealed lots of primary haematite and no pyrite or any pyrite precursor. In fact, the green cherts are clearly the product of diagenesis. As Noble and Renne (1990) conclude, the bottom of the ocean in which the Dekkas cherts were laid down was well oxygenated.

The age of the Dekkas was given as early Guadalupian by Coogan (1960) on the basis of the presence of *Parafusulina californica* and some productive brachiopod species (see Coogan, 1960, p. 255–257). Noble and Renne (1990) report *Albeilella levis*, known from the Middle and Upper Guadalupian in Japan. They also found radiolaria known from the late Permian of both the Quinn River Formation and Japan. The Dekkas passes conformably into the Triassic Pitt Formation. It thus seems likely that the Dekkas spans the entire Middle to Upper Permian and represents a well-aerated ocean bottom in the Panthalassa.

24. The Maitai Group, New Zealand, South Island (Krull et al., 2000). Another "Panthalassan" record has been described from the deposits of an ~400-m-deep basin within a volcanic arc-related basin in New Zealand. The deposit has been described as the "Permo-Triassic Maitai Group" consisting of 16 formations. Only three of them, namely the Tramway, Little Ben Sandstone, and Greville bracket the Permian-Triassic boundary. The Tramway Formation consists of greenish gray sandstones commonly interbedded with darker gray siltstones and carbonaceous claystones. It is a turbidite with pyrite euhedra and complete *Athomodesma* bivalves. On the basis of these bivalves it is dated to the Tatarian (Capitanian to Changhsingian). The Little Ben Sandstone is not datable, but it is gradational with the underlying Tramway. It contains much more volcaniclastics. The Greville is formed from green and gray laminated siltstones, sandstones, and claystones following the Little Ben gradationally. Its age is early Triassic beginning with the Dienerian, i.e., late Induan. In this section, organic carbon content is high in the Tramway, fairly abruptly drops toward the top of the Little Ben, and increases again in the Triassic Greville. Krull et al. (2000) rightly argue that the anoxia seen in this basin cannot be a part of the alleged global anoxia affecting the entire Panthalassan oceanic realm because where it should be highest according to the global anoxia scheme, the total organic carbon in the Maitai Group plunges to a minimum.

25. Arrow Rocks, Whangora Bay, North Island, New Zealand (Takemura et al., 2002). A section has been measured very carefully in this part of Panthalassan deposits in terms of "units" numbered from bottom to top, which we quote in full below (Takemura et al., 2002, p. 292):

Unit 1: 31.5+ m spilitic basalt with four major limestone layers (red: Upper Middle Permian)

Unit 2: 11 m red siliceous mudstone overlain by red, gray, and yellow bedded radiolarian chert (bedding 20–40 cm) (Upper Permian). Fault contact with excision of some strata.

Unit 3: 1 m alternation of chert and black mudstone with minor thin tuff (bedding 1–10 cm) (light gray)

Unit 4: 6.6 m red siliceous mudstone and red radiolarian chert with manganese-rich layers and some tuffs (bedding 5–20 cm). Includes a 1.4 m thick zone of sheared gray chert at the top.

Unit 5: 17.2 m red siliceous mudstone with thin red radiolarian chert (bedding 10–30 cm)

Unit 6: 29.5 m maroon chert and siliceous mudstone (bedding 10–30 cm)

Unit 7: 11.2 m alternation of maroon and green siliceous mudstone (bedding 10–30 cm)

Unit 8: 27 m green siliceous mudstone with vitric tuffs (bedding 20–50 cm)

This section shows neither radiolarian extinction at the Permo-Triassic boundary nor any serious anoxia at any interval as Baud et al. (2006) pointed out. This is another section that shows that the *Panthalassa could not have been anoxic as a whole at any time during the Permo-Triassic.*

CHAPTER VII

History of global redox conditions in the late Permian

In the late Permian world, a number of factors seem to have conspired to reduce the aeration of the seas. The evening of the global climate as a consequence of the end of the late Paleozoic ice age raised global surface temperatures, leading to a warming of the surface waters and thus a diminishing of their capacity to hold oxygen. This poorly oxygenated warm surface water must have made the global oceanic circulation more sluggish than during ice ages, reducing further what little aeration to the deep ocean it otherwise could have contributed. Therefore both the low oxygen content of the late Permian atmosphere and the increased surface temperature of its ocean predisposed the Permian world ocean to poor aeration.

In the previous chapter, we have shown that anoxia in the Permo-Triassic world began developing as a serious oceanic event first in the Paleo-Tethyan deep waters, in places possibly as early as the Carboniferous (Fig. 23). The upper surface of anoxic waters gradually rose to progressively shallower depths in the Paleo-Tethys until the suffocating waters finally invaded the shelves (Fig. 23). Before that, in Kungurian time, it had spilled over into the European rift-controlled seas, possibly helping anoxia indirectly to develop in places as far away as the northeastern Panthalassan marginal basins, scrappy remnants of which now lie embedded in the North American Cordillera. Paleo-Tethyan anoxic waters also spilled out, beginning in the Capitanian, into the equatorial western Panthalassa through the bathyal north Thailand strait(s), polluting certain places directly, but there is now firm evidence showing that there can be no talk of the entire Panthalassa turning anoxic. It in fact seems difficult to keep a water body the size of the end-Paleozoic Panthalassa under the influence of the possible temperature gradients of the late Permian world (see, e.g., Crowley, 1994) from advecting and convecting. The extensive Panthalassan phosphorite deposits bear witness to this, as opposed to the nearly nonexistent Paleo-Tethyan phosphorites of late Permian age (Trappe, 1994; see Fig. 24) most likely because there was no upwelling in a Paleo-Tethys that could not overturn and also because widespread anoxia in the Paleo-Tethys was not favorable to the development of phosphorites, as discussed in the caption for Figure 24.

How badly was the Permian world ocean aerated? This is almost impossible to tell because no ocean floor of that age survives intact and what little record of it has been preserved is found in subduction-accretion complexes containing abyssal deposits in a badly mangled state. Statements made to the affect that the entire world ocean had turned anoxic are based on a few sections found in a few subduction-accretion complexes along Pangean margins, and they are valid only as far as

those sections are concerned because we cannot tell from the difference between their ages of origin and ages of incorporation into subduction-accretion complexes how far into the Panthalassa they had been located (as, for example, Isozaki, 1997, had attempted) because we have no evidence of the direction and uniformity of the plate motions responsible for carrying them into their final destinations. Even if we were able to tell where those sections came from, it still would not have been possible to generalize to an entire ocean from a few sections as a map of present day oceanic anoxia clearly shows (Fig. 20). Imagine a geologist studying the records of today millions of years hence: what would she or he conclude should she or he have at hand only the anoxic areas offshore the Gulf of Mexico, offshore Venezuela, and offshore Makran? Would she or he be justified to conclude that there was widespread anoxia in the oceans in the Holocene because such widely separated records show the same thing? Certainly not. Such a person could, however, develop a hypothesis to that effect and start looking for a deep ocean record preserving an oxygenated deep-sea record to test her or his idea. In case she or he succeeds in finding even just one such record, that would be enough to refute the earlier hypothesis that the entire Holocene ocean had been anoxic. We believe we have just done that for the late Permian ocean: we found sections in which there is no substantial anoxic record in the late Permian and no serious record of local extinction.

Lacking complete coverage, we need other sorts of indirect evidence to assess the degree and extent of anoxia in former ocean basins. We have developed a methodology in following the temporal evolution of anoxia in different depth intervals in the world ocean (Fig. 23). We looked at anoxia at ocean bottoms, at distal, i.e., topographically lower, shelves, and, finally, in the proximal, i.e., topographically higher, shelves. We found independent criteria to assess the elevation of the shelves we studied and established that there was a definite pattern in the spread of anoxia from the deepest ocean to the shallowest parts in the fairly limited volume (with respect to the rest of the world ocean) Paleo-Tethys. No such evolution is seen anywhere else. It is unlikely that anoxia developing on individual shelves, owing to local development of restricted conditions, will everywhere follow the same pattern. There are, for example, latest Permian oxidized sequences in areas such as the Shahreza section in Iran (Heydari et al., 2008), but that section displays a very shallow water sequence on a carbonate platform facing the opening Neo-Tethys and on mobile fault blocks similar to the situation in the Greenland-Norwegian rift system. Therefore, any section that is anomalous must be carefully scrutinized in terms of its geological setting and only then can one make a statement about its message concerning the overall redox conditions in the Tethyan realm or in the Panthalassa. We have looked at enough shelves to assure ourselves that the pattern is probably general within the Paleo-Tethys, although this is still a working hypothesis that can use more observations. *A single contrary observation, if inexplicable owing to local unusual conditions, will refute our hypothesis of the progressively rising surface of the anoxic waters in the Paleo-Tethys definitively, and therein lies its great explanatory power.*

Figure 24. The "killing fields of the Paleo-Tethys." Global-fungal-remain localities indicating the areas of maximum amounts of organic debris resulting from dead bodies perished during the killing event at the end of the Permian. Data are from Steiner et al. (2003) and Visscher et al. (1996). This event we interpret to have been an eruption of the Paleo-Tethys (H_2S and ?CH_4) as all the reported localities form a halo around it and along its deep gulfs (Malagasy gulf) and its spillways (Norwegian-Greenland rift system and north Thailand strait[s]). The width of the areas of fungal distribution is compatible with the area expected to be affected during the eruption of the Paleo-Tethyan ocean. The numbers refer to the following localities: 1—north Alaska; 2—Sverdrup basin, Arctic Canada; 3—Svalbard; 4—Barents Sea; 5—Tunguska basin, Siberia; 6—Pechora basin, northern Russia; 7—east Greenland; 8—British Isles; 9—North Sea; 10—Zechstein basin, Germany; 11—Zechstein basin, Poland; 12—Moscow basin, Russia; 13—Mangyshlak, Kazakhstan; 14—Transdanubian Mountains, Hungary; 15—Southern Alps, Italy; 16—Dinarides, Bosnia; 17—Tunisia; 18—Negev, Israel; 19—South Anatolia, Turkey; 20—Saudi Arabia; 21—Sichuan, south China; 22—Meishan, south China; 23—Salt Range, Pakistan; 24—Mombasa basin, Kenya; 25—Morondava basin, Malagasy; 26—Raniganj basin, India; 27—Bonaparte Gulf basin, Western Australia; 28—Banda Sea; 29—Bowen basin, Queensland; 30—Sydney basin, New South Wales; 31—Karoo, South Africa. The green circles are marine phosphorite localities, except that those with an inscribed letter "L" are of lacustrine origin. The large green circle is the only currently mined major phosphorite deposit of late Permian age. All phosphorite information is after Trappe (1994). The absence of major phosphorite deposits despite the abundance of organic nutrients around the Paleo-Tethys most likely indicates widespread anoxia and the paucity of oxygen, necessary for the formation of the phosphorites (cf. Baturin, 1971; Franquin et al., 2006). Burnett (1977) demonstrated that marine phosphogenesis in the upwelling region along the modern continental margins of Peru and Chile is most intense at the edges of the oxygen minimum zone where the latter intersects the seafloor, rather than in the areas of lowest oxygen levels in bottom waters. Like-

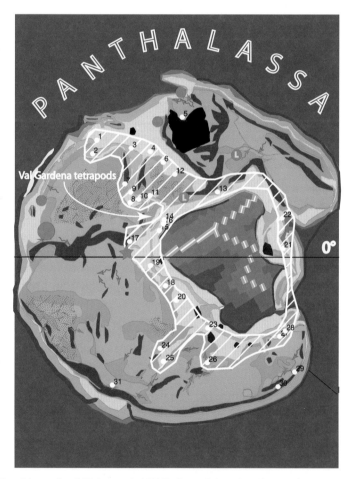

wise, in the oxic, non-upwelling region on the eastern continental margin of Australia, O'Brien et al. (1990) showed that phosphogenesis occurs during early burial diagenesis within bottom sediments at oxic and/or anoxic interfaces; the phosphate particles which thus formed later become concentrated into phosphorites through reworking by vigorous, oxygenated bottom currents. Both the South American and Australian occurrences indicate that the major processes promoting phosphogenesis and the formation of phosphorites require some oxygen (Kolodny, 1981; Jarvis et al., 1994), which most likely was not available in sufficient quantities in the deep waters of the Paleo-Tethys. The lone green star is the only peri-Tethyan late Permian phosphorite locality. The phosphorite occurs in a fine-grained sandstone and is of no economic value. Although Parrish and Ziegler (1983) report the occurrence of Permian phosphorites in the Hazara District of Pakistan, on the basis of Bhatti's 1977 report, that deposit is in reality of Cambrian age (Bender and Raza, 1995, p. 270).

CHAPTER VIII

Paleo-Tethyan anoxia and the "end-Permian" extinction

If the Paleo-Tethys indeed became gradually anoxic, what consequences might this have had on the inhabitants of our planet? One obvious result would have been extinguishing most aerobic life in the Paleo-Tethys itself. If it also erupted and spewed its poisonous gases into the air, it would have killed most aerobic life around it within a killing halo of not less than a 2000 km. width, if scaled to the present-day lake eruption events (Zhang and Kling, 2006). The Paleo-Tethys would have done all this, at a time when, so paleontology teaches us, indeed nearly 95% of all life, marine and terrestrial, really *did* become extinct on our planet. Is this just a coincidence or was Paleo-Tethys really the culprit?

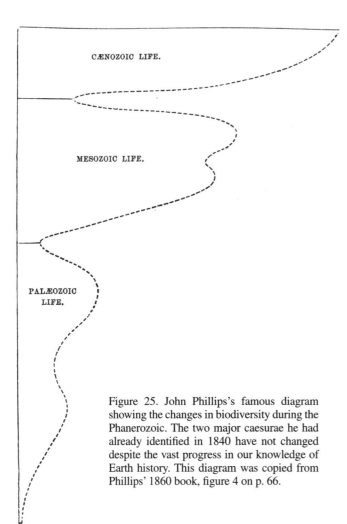

Figure 25. John Phillips's famous diagram showing the changes in biodiversity during the Phanerozoic. The two major caesurae he had already identified in 1840 have not changed despite the vast progress in our knowledge of Earth history. This diagram was copied from Phillips' 1860 book, figure 4 on p. 66.

The Problem of Extinctions

Ever since Georges Cuvier showed that the mammoths and wooly rhinoceroses, and with them an entire fauna "before the empire of man," had become extinct (Cuvier and Geoffroy, 1795; Cuvier, 1796, 1812, p. 1–116),[37] and John Phillips showed that twice in the Phanerozoic the biodiversity had plunged into appalling minima and used these to subdivide the Phanerozoic into three major eras (Phillips, 1840, 1860; Fig. 25), and Alcide d'Orbigny (1852) used the extermination of entire faunas to define what we today call stratigraphic systems, in which endeavor he was followed by James D. Dana (1863, p. 397–399), causes of what has been called "universal extinctions" (Dana, 1863, p. 398), "massive extinctions" (Bramlette, 1965, p. 1696), "mass extinctions" (Newell, 1967, p. 69), or "diversity crashes" (e.g., Slipper, 2005) have bothered earth and life scientists alike (Suess, 1860, 1909a, p. 739–782, 1909b; Dacqué, 1936, p. 145–152; Peters, 1949; Schindewolf, 1950, esp. p. 92–108 and fig. 17, 1954, 1962; Beuerlen, 1962, 1975, p. 311; Newell, 1967; Alvarez et al., 1980; Hsü, 1980, 1987; Raup and Sepkoski, 1982; Sepkoski, 1986; Hallam, 1987; Officer et al., 1987; Prothero, 1994; Kauffman and Erwin, 1995; Archibald, 1996; McGhee, 1996; Seilacher, 1999; Adatte, 2003; Foote, 2003; Bambach et al., 2004; Purdy, 2007; for general reviews, see Hsü, 1986; Raup, 1986; Stanley, 1987; Donovan, 1989; Erwin, 1993, 2006; Ward, 1994, 2000, 2006; Officer and Page, 1996; Alvarez, 1997; Hallam and Wignall, 1997; Courtillot, 1999; Benton, 2003; Buffetaut, 2003; Ellis, 2004; Taylor, 2004; also see the conference report by Kauffman and Walliser, 1988). Since Cuvier's epoch-making 1812 book, there has been little doubt that entire faunas did indeed disappear in the geological past, but ever since Lyell (1833, esp. p. 26–34), it has remained contentious as to how pervasively and how quickly these events occurred. Arguments concerning the nature and origin of these caesurae in the progression of life have long been plagued by

---

[37] It is almost universally claimed that Baron Georges Cuvier was the first to show, on the basis of comparative anatomy, the extinction of a former species. This is not true; the first person to do so was a medical student in Germany, Johann Christian Rosenmüller (1771–1820), in his dissertation, entitled *Quaedam de ossibus fossilibus animalis cuiusdam, historiam eius et cognitionem accuratiorem illustrantia*, presented to the University of Leipzig in 1794. In this thesis Rosenmüller identified the fossil cave bear as a separate species and named it, according to the rules of the Linnean binomial classification, *Ursus spelaeus*. This was the first time a new species was identified on the basis of comparative anatomy performed on fossil material. A year later Rosenmüller published a German version of his thesis (Rosenmüller, 1795). For a remarkable history of this discovery and Rosenmüller's work, see Rosendahl et al. (2005). However, it is true that it was Cuvier's work that made the method and its usefulness famous worldwide. We thank Dr. Robert Darga, the director of the Siegsdorf Naturkunde- und Mammut-Museum in Bavaria, Germany, for drawing our attention to Rosenmüller and his work and for providing us with the relevant literature.

uncertainties as to what constitutes a mass extinction and which mass extinction is greater or lesser than any other. For instance, the end Permian extinction has long been regarded as the greatest of all extinctions during the Phanerozoic, "a nearly complete destruction of all life" (Raup, 1986, p. 49), a time "when life nearly died" (Benton, 2003). However, not all "mass extinctions" are alike with respect to the effect they had on the biomass of the planet: some threatened all life, while others only a part of it; some came after a long prelude, while others were sudden.

Figure 26 shows the segment of Purdy's (2007, fig. 1) biodiversity curve between 450 Ma, i.e., the great Ordovician biodiversification event (Schindewolf, 1954, Table 3; Webby et al., 2004; Webby, 2004, fig. 1), representing the maximum booming of Sepkoski's (1981) "Paleozoic fauna," and the present, plus two least-squares-fitting "trend-illustrating" curves and the times of eight "mass" extinctions (Şengör et al., 2008). Şengör et al. (2008) argue that between the present and 450 Ma ago, the total genera curve displays some striking features. The most significant is that the diversity minimum at the end of the Permian not only defines a colossal plunge, but it also divides the time series into two parts with very different characteristics. From around 450 Ma up to the Permian extinction, the curve displays a steady oscillation with the main peaks of biodiversity decreasing slowly, although the short-lived end-Silurian extinction had the greatest rate of reduction in biodiversity after the K-T extinction! The oscillation is not periodic, strictly speaking, and does not present any erratic behavior. The four main peaks are separated from each other by periods of around 50–70 million years (average duration of a collisional orogeny and about half of the life-span of a typical Phanerozoic rifted margin: see Bradley, 2007) and the oscillation has a significant variance around the mean. This is

almost certainly a driven oscillator that has the frequency characteristics of the force that drives it, because the frequency structure of the extinction intervals between 450 and 250 Ma ago is very distinct from the pre-450 Ma ago structure and from the structure that succeeded it (Fig. 26). The two biodiversity curve structures preceding and succeeding Sepkoski's (1981) "Paleozoic fauna" biodiversity curve structure are in turn very different from one another. Thus, these cannot be the natural modes of the oscillations of the biospheric diversity, at least not any two of these. It is thus very likely *that whatever caused the oscillations in the Paleozoic was most likely a common mechanism or a group of mechanisms with common characters outside the biosphere (with the end-Silurian exception?).* However, when a new evolutionary episode begins (i.e., a certain period of time in which the genera time series manifests a behavior distinct from the others, especially the one that predates it), the sensitivity of the system on this new "initial" condition might be different from the sensitivity of the previous system on its own initial condition. Characteristics such as the proportionalities of the birth-death processes of different species belonging to the same episode, speeds, and magnitudes of disease propagations, etc., are also likely to be different. Therefore, it is difficult to make a one-to-one analysis on the states of systems, which, at certain periods of their reign, might include exactly the same number of genera. The left-hand (post-Permian) side of the curve in Figure 26, on the other hand, can be represented by an almost linear fit, save for the sudden plunge that corresponds to the K-T decay and a comparatively large decay around 210 Ma, that is, the first biodiversity plunge after the recovery from the Permian extinction. The post-Permian part of the curve has a small variance around an almost linearly increasing trend, apart from the K-T decay that clearly has an origin different from all the rest. If we artificially make the amplitude of the late Permian reduction in biodiversity similar to those of the preceding extinction events after the Cambrian, we can then take the late Triassic–early Jurassic extinction event as the point dividing the Ordovician to Recent time series into two parts with entirely dissimilar characteristics. However, this was manifestly not the case and the late Permian extinction was almost twice as effective in reducing biodiversity as all the preceding post-Ordovician extinctions and the succeeding end-Triassic extinction episode. However, contrary to the conclusions of Bambach et al. (2004), the causes of the extinctions from 450 Ma ago to 210 Ma ago were probably similar *with an additional cause during the end-Permian extinction that increased its magnitude* (Şengör et al., 2008).

The "End-Permian" Extinction

The "end-Permian" extinction has long been thought so severe to be unique in the Phanerozoic. The severity, however, has never been properly defined, but assumed to be prodigious because of the abysmal level to which the terrestrial biodiversity had sunk by Induan time. The *rate of biodiversity diminution* was, however, much less than that during the end-Cretaceous

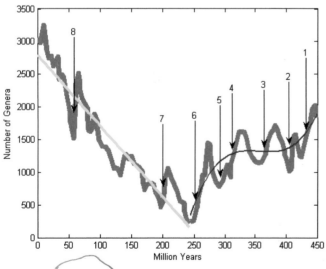

Figure 26. Biodiversity change through the Phanerozoic simplified from Purdy (2007, fig. 1) showing the major extinctions and least-squares fittings to show trends of the changes in biodiversity since the great Ordovician biodiversification event (from Şengör et al., 2008).

extinction, probably setting these two global events apart from one another in terms of their causative mechanisms, and Şengör et al. (2008) have shown, on their newly-defined G-scale of extinctions, that the G (greatness) number of the Permian extinction was actually smaller than that at the end of the Cretaceous. Which processes acted in such a way as to reduce biodiversity during the Permian then? To answer that question, we must outline both the temporal and the spatial characteristics of the Permian extinction and the *common biology* of both the victims *and* the survivors.

Who Died?

The best short statement concerning the temporal evolution of both the marine and the terrestrial extinctions at the end of the Paleozoic remains that by Steve Stanley, stated in his wonderful, popular book of 1987, despite all the superb and detailed research undertaken since. Stanley (1987, p. 97) pointed out that great transitions had taken place on land and at sea and that these were not confined to the very end of the Permian, but spread over a time interval of some 10 million years and perhaps more. This remains true (the time interval is too long to be an artifact of the Signor-Lipps[38] effect of preservation), and the time interval becomes ever longer as more research results accrue, effectively disposing of explanations involving fast events such as meteorite impact and catastrophic vulcanicity confined to a million years or less.

Stanley's later study with Yang Xiangning echoed an earlier conclusion by Heinz Kozur that indeed there were not one but two peaks of extinction in the marine biosphere: one at the end of the Guadalupian and the other at the end of the Changhsingian (Kozur, 1977, 1980; Stanley and Yang, 1994; Yin et al., 1994), with 58% of all marine genera disappearing at the end of the Guadalupian (Stanley and Yang, 1994) and anywhere between 83% (Sepkoski, 1989, 1990) and 96% (Raup, 1991) disappearing at the end of the Changhsingian.

The groups affected in the Guadalupian extinction were mainly benthic organisms including rugose corals, bryozoans, fusulinid foraminifera, and articulate brachiopods plus the necto-benthic ammonoids, and "the extinctions happened especially in Tethyan, mid-latitude regions; there is very little evidence for the event in northern waters" (Benton, 2003, p. 256). Powers and Bottjer (2007) recently added the remarkable observation in their compilation of bryozoan data that extinctions not only started in the Capitanian, but initially in the deep sea and from there progressively invaded shallower areas, leading them to suggest that they were a result of "environmental stress result[ing] from the gradual encroachment of some deep-water phenomenon onto the shelves" (*ibid.*, p. 995). In areas only irregularly, or not at all, affected by Paleo-Tethyan anoxia, bryozoans continued living on happily as in Greenland and Spitsbergen (Nakrem, 1994). Therefore, in the photic zone of the Paleo-Tethys, the extinction happened abruptly in the latest Permian. Farabegoli et al. (2007) found a similar pattern of

the benthic taxa in the Southern Alps: they suggest a diachronous disappearance of the bentich community in the deep waters of the Paleo-Tethys, where the higher shelf of the Southern Alps acted, for a time, as an asylum[39] for the survivors.

But this is precisely how anoxia developed at the same time and in the same place, i.e., in the vast, closed equatorial ocean, the Paleo-Tethys, throughout the Permian, but it reached different realms of life such as the abyssal plains, the bathyal slopes, the lower shelves, and the higher shelves at different times, as the data summarized in this book document and as Wu et al. (2007) recently also found. When Kozur (1977) looked at the entire Permo-Triassic extinction picture, he, too, was struck by the fact that fusulinids, trilobites, tabulate corals, some 90% of the brachiopod families, most of the bryozoans, bivalves, gastropod genera, crinoids, and blastoids, representing mostly the warm water benthos, had disappeared whereas the colder water fauna was not affected. For example, three ammonoid genera survived the Permo-Triassic extinction (Yang and Wang, 1999) and they were all cooler water forms not living in the Paleo-Tethys. Indeed Brayard et al.'s (2007) maps illustrating the biogeography of the ammonite recovery between the Griesbachian and the Spathian show not one ammonite genus living in the Paleo-Tethys and that Triassic recovery developed along the easternmost "threshold" areas over the Cathaysian bridge and across the vast expanse of the Panthalassa. The ammonites ventured across the Paleo-Tethys earliest during the Smithian and began colonizing European shores (see fig. 8 in Brayard et al., 2007). Wu (2006) repeats the same general observation that the taxa that went extinct were warm water, stenobiontic organisms, whereas the few survivors were eurytropic and eurthermal organisms inhabiting cold water environments. Also, ~50% of the genera that had disappeared from the record reappeared in the Olenekian to Ladinian interval forming Lazarus taxa. They, thus, must have found asylums somewhere not now preserved. We suggest that they may have gone to the Panthalassa, using not only the north Thailand strait(s), but also some of the shallow epicontinental marine connections across the Cathaysian bridge to escape from the gas chambers of the Paleo-Tethys. What better place than an ocean occupying more than half the world and whose record today is all but gone

---

[38]See Signor and Lipps (1982). This effect is claimed to result from infrequent preservation of fossils. Because of sporadic preservation, members of a particular taxon in different localities may appear to have become extinct before the actual "extinction time" of that particular taxon, and some think that the Signor-Lipps effect behooves geologists to take the time of the latest occurrence as the time of catastrophic universal extinction of the taxon. While not unreasonable in cases of short durations of "gradual extinctions" (at most a couple of millions of years), this method of fixing the time of extinctions has severe logical and epistemological problems identical with those attending Hans Stille's "indirect method of dating orogenies." For Şengör's criticisms of this method and its consequences, see Şengör (1991c, especially p. 440), and for the late, regretted Bill Sarjeant's recommendation of this paper to catastrophist paleontologists, see Sarjeant (1992). In fairness to Phil Signor and Jere Lipps, however, we must underline that they wrote their paper to sound a well-justified cautionary call and not to state a dogmatic rule as some seem to have taken it since.

[39]We here use Eduard Suess's (1909) word for what has since the fifties of the last century been called refugia, because the priority of the concept belongs to Suess.

because of subduction, a process no one even dreamed of before plate tectonics, at scales we now think not only possible, but necessary. The widely quoted peculiarity of the Permian-Triassic extinction is that in the earliest Triassic there was a rebound of life and then another but more restricted wave of extinction. Both the rebound and the re-extinction are easy to explain by the refilling of niches from asylums and then an increased struggle in a disappearing realm, such as the Paleo-Tethys was.

Hallam and Wignall (1997, p. 116) express puzzlement as to why nectonic and/or necto-benthic creatures should have escaped the extinction. For instance, fish groups experienced no extinction or only insignificant extinction (Schaeffer, 1973; Patterson and Smith, 1987). It must be because they could swim away from the polluted areas and because there was a place to run to that was very large, big enough to accommodate them all. The only benthic groups that could weather the crisis were all dysaerobic organisms of the Paleozoic (Hallam and Wignall, 1997, p. 116). For example, amongst the brachiopods, the highly provincial Changhsingian faunas were replaced by the widespread dysaerobic fauna such as *Lingula* sp. (Kammer et al., 1986), the champion survivor of all times.

We just mentioned that many families of foraminifera and the suborder Fusilina had vanished (Tappan and Loeblich, 1988). They disappeared shortly *before* the end of the Permian (Kozur, 1998). At the end of the Guadalupian, around 45 genera (larger, more complex fusulinids, Ross and Ross, 1995) went extinct. After this extinction many foraminiferal lineages radiated in the Lopingian until the more severe extinction at the latest Changhsingian (Altıner, 1981; Pasini, 1985; Noé, 1988; Tong, 1993; Baud et al., 2005; Groves and Altıner, 2005), and it mostly affected the tropical, more complex forms (Brasier, 1988). The survivors were mostly dysaerobic groups in this case too.

All in all, the available paleontological evidence shows that marine extinctions began in the Paleo-Tethys already in the medial Permian, in the lowest parts of the ocean amongst those attached to the ocean floor. With time those living in higher and higher places became affected. Those who could swim largely got away. Finally, outside the Paleo-Tethys, not much really happened; but then, there was not much life space outside the Paleo-Tethys except the vast abyssal plains, and the endless expanses of the surface, of the immense Panthalassa, neither of which were much favored by organisms except the radiolaria, whose record often gets subducted (cf. Moore, 1975). Because of the circum-Pangean subduction, shelf space, the prime abode of marine life, along the Panthalassan margins was limited and constructed mostly on tectonically and sedimentologically unstable forearc areas. All this indicates that it was the poisonous gases in the Paleo-Tethyan anoxic waters that killed the inhabitants.

The geographical confinement to the Paleo-Tethys further disposes of truly "global" causes, such as the alleged superanoxia events poisoning the entire world ocean. Moreover, if a flood basalt province close to the Arctic (in late Permian time), as the Tunguska Province was, leaves no record of killing in its vicinity, but somehow concentrates most of its killing power in an isolated ocean far away, its culpability in the Permian extinction events would be suspect.[40]

There are temporal and causal parallels in the land biosphere to the events we just reviewed in the marine biosphere. For example, the paleophytic flora, represented by the extensive coal swamp trees such as *Sigillaria,* went out by the end of the Guadalupian and the mesophytic flora, more adapted to a dryer environment (and topographically higher sites!) and composed of such plants as conifers, appeared in the late Permian. The mesophytic flora mostly characterized the dry terrain and the highland areas dominantly beyond 30° north and south latitudes in the later Permian world. Despite that, large trees only appeared as late as late Triassic time (e.g., 60-m giants are known from the Chinle Formation in the western United States that spans an age from the early Carnian to the late Rhaetian: see Fastovsky and Smith, 2004, p. 615). Sparsity of forests populated by trees smaller than both those of the Carboniferous and the late Triassic characterized the low oxygen environment in the late Permian and early Triassic (Fig. 18), yet it was that flora that became the ancestors of our flora today.

Vertebrate life also underwent significant changes. The dominant reptiles changed from the dorsal-finned, mammal-like but necessarily slow-moving Pelycosaurs, with roots in such late Carboniferous forms as *Ianthosaurus* sp., to the somewhat more agile Therapsids, the nonmammalian taxa of which survived into the early Cretaceous (McLaughlin, 1980; Carroll, 1988, ch. 17)[41] only to be gradually replaced by the highly agile dinosaurs, the flying Pterosaurs that had began evolving in the Ladinian (Desmond, 1977; Carroll, 1988, chs. 14, 15, and 16; Benton, 2004) and finally by the true mammals in the Norian (Carroll, 1988, ch. 18). The real change is often said to have happened from the "paleotetrapods" to "mesotetrapods" in the Carnian to Norian interval (Charig, 1984; Benton, 2004).

We do not share this view. The beginning of the Capitanian or perhaps even the end of the Wordian saw the end of the Dinocephalians (Spencer G. Lucas, 2007, personal commun.[42]), the largest tetrapods of the Paleozoic world, and they were replaced by the generally smaller, essentially mammalian Gorgonopsians (Sigogneau, 1970; Battail, 2000; Gebauer, 2007) and Dicynodonts (King, 1988; Battail, 2000). This transition to a more mammalian-like body was, we believe, the decisive change in the tetrapod world, and its earliest heralds should perhaps be placed already into the end of the Carboniferous, into

---

[40]We note the verdict of Self et al. (1997, p. 406): "The most reasonable statement, given current knowledge, is that a continental flood basalt eruption probably could not cause mass extinctions, but a series of them during the growth of a CFB [continental flood basalt] province would have been able to stress the environment to such an extent that any other major perturbation would have had a more extreme effect."

[41]The Pelycosaur-Therapsid changeover happened across what Lucas and Heckert (2001) called "Olson's gap," spanning an interval from uppermost Kungurian to bottom Kazanian (Upper Cisuralian to uppermost Roadian) in which we have no record of Permian tetrapods anywhere (Lucas, 2004).

[42]Dr. Lucas writes that his interpretation of the timing of the Dinocephalian extinction is a minority view opposed to the majority view of a late Capitanian extinction.

the dawn of the amniote world, as already implied by Gregory in 1955 and as is consistent with Kemp's (2006, p. 485) argument that "relatively constant rate of acquision of mammalian characters" determined the origin of endothermic mammals. The same is certainly true for the dinosaurs. But, after the Capitanian, cold-blooded tetrapods never again ruled the world; instead the "mammalian plan," i.e., the synapsids, and the "bird plan," i.e., the diapsids, became the two dominant forms of the amniotes, namely the theropsida (beast-faced) and the sauropsida (lizard-faced) (Goodrich, 1916), the latter by far predominating over the former during the Mesozoic, although it originated later (see Carroll, 1988, ch. 10). At the end of the Lopingian, the Gorgonopsians also disappeared and only two or three Dicynodont species plus some small diapsids such as the *Proterosuchidae* (Benton, 2004) survived the Permo-Triassic transition.

What Were the Characteristics of Those Who Survived?

They were small (Price-Lloyd and Twitchett, 2002), lowland, partly aquatic tetrapods (Shiskhkin, 1997) that could not traverse even moderately high mountains such as even the lowest parts of the Appalachians and the post-Hercynian Basin-and-Range-like environment of Europe, judging from their distribution. They were adapted to live in the low-oxygen world of the end-Permian and for that reason seem to have preferred the relatively lower and less harsh eastern Pangea as opposed to the desert- and mountain-dominated western Pangea as their abode. The *Lystrosaurus*, for example, originated very near the Gulf of Malagasy of Gondwana-Land, in east Africa (King and Jenkins, 1997), and marched east (and southwest) and most likely headed north across the Cathaysian bridge, in the footsteps of its *Dicynodon* cousins (Battail, 1997), to reach the extensional lacustrine basins of the post-Altaid world in central Asia,[43] from where it made it to the lowlands of the Moscow basin in the essentially depopulated world of the early Triassic.

Recently a very similar scenario for the paleobiogeography of the edopoid amphibians has been proposed by Steyer et al. (2006). They showed that these animals, mostly confined to areas far from the Paleo-Tethyan shores after the late Carboniferous, were able to cross the mountains of western Pangea until the late Permian, when their world disintegrated, because now the mountains functioned as barriers. But the late Permian mountains of western Pangea must have been at least as low if not lower (because of widespread extension affecting them) than those of the earlier Permian. Steyer et al. (2006) hold global warming responsible for making the mountains into barriers. This is indeed most likely for amphibians because it would imply less water on the way, but the air they had to breathe was also becoming less oxygenated, most unsuitable for these lowland creatures. The story edopoid amphibians tell us is thus similar to that of the Dicynodonts.

But the world was rapidly recovering. The archosaurs (ruling reptiles) soon produced prodigious predators (Benton, 2004, fig. 1.9) and the unassuming *Lystrosauri* disappeared, probably mostly into the bellies of the early archosaurs at the latest by late Olenekian time (Damiani et al., 2000). But the primitive archosaurs could not keep pace with their much more sophisticated cousins who dominated the world from the late Anisian onwards to the end of the Cretaceous. Throughout this evolution, two things increased and in the following order: agility and then size, both probably products of better respiration. It is not a coincidence that Archosauria produced its paleotetrapod-like lineages during the early Triassic coal gap and its most agile and dominant lineages in the "bird plan" almost simultaneously with the evolution of large trees after the coal gap had passed in the early Anisian (cf. Benton, 2004, fig. 9.1; Retallack et al., 1996). Further tetrapod evolution has been dominated by dinosaurs whose initial slow evolution seems to have resulted from the small degree of intra- and intercontinental habitat fragmentation and intra- and interordinal competition (Sereno, 1999), but to this day they (now represented exclusively by the birds, i.e., avian dinosaurs, with approximately 8,500 extant species) remain the dominant land vertebrates. The mammal body plan (with some 5400 extant species) seems not nearly as successful as the bird body plan. The dinosaurs appear to have achieved this success after the end-Cretaceous extinction by proliferating small to medium size species and making flight their preferred mode of locomotion (the most successful mammals with 20% of all mammal species, the bats, are also avian!).

The picture we obtain from the land biosphere is that the Guadalupian saw a first wave of extinction and the end of the Lopingian another one, possibly one that gradually accelerated through the late Permian to the final catastrophe, leaving the Triassic world with very few genera to begin with (e.g., Ward et al., 2005). Both extinctions worked to eliminate large animals requiring agility. Almost no carnivores were left by the end of the Permian (exceptions include the possibly venomous Eutherocephalia*ns*), despite the fact that seemingly many succulent *Lystrosauri* were still wandering about. The progressively more mammal- and avian-like bodies require more and more oxygen than other tetrapods and any diminution in oxygen would first eliminate those that are most mammal- or bird-like and, among them, the larger ones that require the most energy. This remains true, even when one reviews the osteological record with the caution rightly recommended by Angielczyk and Walsh (2008). As has long been known and as Seilacher (1999) again pointed out recently, extinctions always eliminate the most specialized forms. However, the immense advantages of the mammalian and avian ability to remain energetic for hours must have outweighed any other advantage, because, after every extinction, the surviving lineages created more and more mammal-and avian-like organisms in the amniote world.

Even insects suffered terribly. Labandeira and Sepkoski (1993) and Labandeira (2005) pointed out that the only mass extinction insects suffered came at the end of the Permian. There

---

[43]For the development of the basins, see Allen et al. (1995) and Wartes et al. (2002). For the *Lystrosaurus* finds in central Asia, see Yuan and Young (1934) and Young (1935, 1939).

are twenty-two orders in the insect class (out of 37 total) known from the Permian and eight of them disappeared with another five losing many families by the end of the period. Only one or two insect orders vanished since that time.

The Mechanism of Killing

What is it that is common to plants, tetrapods, and insects living on land and makes life possible for them? The only thing we can think of is the air they breathe. If that air gets poisoned they will die. This is something they seem to have shared with their counterparts living in the Paleo-Tethys. How could that happen? How could the internal dynamics of an ocean kill not only organisms in it, but also around it? The only mechanism capable of doing this that we know of is gas eruption in a manner suggested by Zhang and Kling (2006), Ryskin (2003), and Kump et al. (2005). If the gases that resulted from anoxia in the ocean end up erupting, they would create a field of devastation around the ocean commensurate with the volume of gas released.

At this point the observations by Payne et al. (2007) concerning the existence of erosion surfaces on uppermost Permian skeletal limestones in south China, Turkey, and Japan, and their inference based on sedimentary facies, microfabrics, carbon isotopes, and cements that this erosion occurred in a *submarine setting possibly because of rapid carbon release from sedimentary reservoirs or the deep ocean,* corroborates the idea that massive gas release may well have occurred at least within the Paleo-Tethyan and some nearby polluted Panthalassan shelves. This is the only piece of possible direct evidence of gas eruption in the Paleo-Tethys that we are aware of. Could a similar, but subaerial erosion with the absence of any indicators of normal subaerial processes reported from Guizhou grand bank by Collin et al. (2008) be an indication of transient marine erosion owing to tsunami-like, sea-level raising waves created in a bubble-agitated Paleo-Tethys? Whatever the cause of the erosive surfaces reported by Collin et al. (2008), they seem related to the events Payne et al. (2007) described.

Another piece of evidence, albeit indirect, gives independent corroboration to the gas eruption hypothesis: Sheldon (2006) found, at the Graphite Peak in Antarctica (85°05′ S, 172°37′ E), that there was an abrupt increase in chemical weathering in the earliest Triassic among otherwise genetically similar paleosols, shown by greater leaching, greater accumulation of immobile rare earth elements (REEs), and evidence for lower pO_2. Because various paleosols seemed in stratigraphic continuity, Sheldon interpreted his observations as indicating a Triassic greenhouse and the presence of methane release during the extinction and its immediate aftermath. Because the Graphite Peak is a part of the Transantarctic Mountains, we cannot now know where it was at the end of the Permian, but it is likely that it was closer to South Africa than it is now (cf. Şengör et al., 2001).

We have no means of judging how much gas Paleo-Tethys released—if any indeed—but we have a pretty good idea as to how effectively it must have killed on land.

Henk Visscher and his colleagues (1996) pointed out that throughout the world, sedimentary organic matter preserved in end-Permian successions contain unparalleled abundances of fungal remains irrespective of depositional environment, be it marine, lacustrine, or fluvial, and irrespective of floral province or climatic zone. They interpreted the fungal abundance as reflecting a massive dieback of arboreous vegetation most likely subsequently affecting the fauna and leading to its collapse as well. When one plots their "worldwide" observations, it is remarkable how tightly they form a killing halo around the Paleo-Tethys and its channel of communication with the Panthalassa between the port of Dobrudja and the Greenland-Norway rift system (Fig. 24). There are isolated reports of "fungal spikes" from the Karoo Province (Steiner et al., 2003) and from western Australia, but nothing quite like the abundance from around the Paleo-Tethys. This, to us, shows that the Paleo-Tethys inflicted death not only inside it, but also around it, in a halo of indiscriminate massacre, sparing neither beast nor vegetable. Although the data are very sparse, still, when we plot the localities of the Platburgian (±Lopingian: Lucas, 2006) tetrapod finds on the same Pangea reconstruction showing the killing fields of the Paleo-Tethys, we notice that during the Lootsbergian (±Induan and part Olenekian: Lucas, 1998) there was hardly any recovery within the former Paleo-Tethyan killing fields; at least there are no earliest Triassic (not counting the latest Permian *Lystrosaurus* zone) tetrapod reports from within the killing fields, despite abundant sedimentary deposits of appropriate age and facies (Fig. 15E). Whether this observation is an illusion caused by the sparse data we have at hand now (Lucas, 1998, 2006) or reflects a real absence, only further diligent search for late Permian and early Triassic tetrapods around the Paleo-Tethys by our vertebrate paleontologist colleagues can reveal.

But can marine gas eruptions in the Paleo-Tethys explain land extinctions in South Africa, so far away from the Tethyan realm? This depends on whether gas eruptions also happened in the Gulf of Malagasy (Fig. 15C) and, if so, how effectively. The extension of the killing fields to the surroundings of this long inlet into the heart of Gondwana-Land (Fig. 24) allows the postulate of an effective poisoning here too, but the observation points are as yet too sparse to be sure. Another intriguing possibility is the presence of large late Permian lakes in South Africa in the eastern Karoo Trough containing horizontally laminated gray mudstones and sandstones. Tankard et al. (1982, p. 390) point out that the sparse red colorations in these sequences cut across rock-type boundaries and are clearly diagenetic in nature and therefore the environment of deposition must have been anoxic. Tabor et al. (2007) show that some of these lakes terminated their lives in "stinking sulfurous swamps." Could these lakes have joined in the gas-eruptive concert? To cause medial to late Permian land extinctions they need not be exactly synchronous with the Paleo-Tethyan eruptions. This is an intriguing hypothesis well worth testing and can be extended into similar lakes farther north in the Congo, which, however, prior to this writing we have not been able to do because of the lack of material available to us.

In short, the land animals went extinct because something happened to the air they were breathing and whatever happened, happened at least twice in the Permian. This is precisely the record that we read from the marine deposits.

This double-phased extinction originated in a geographically well circumscribed area, making single shot global explanations impossible to entertain. This makes bolide impact, Tunguskan traps, global warming, and global eustatic events as equally unlikely candidates for the Permian extinctions. There were other immense trap eruptions, such as the ±1.5 million km³ basalts of the Paraná-Etendeka province, once claimed to have erupted in a time span of less than a million years (e.g., Renne et al., 1992), but now known to have spanned a time period of some 5 Ma (Peate, 1997), and which occupy nearly half the size of the Tunguska traps, that did, however, little to the biosphere and certainly has nothing to do with the Jurassic-Cretaceous boundary (the traps are more than 10 million years younger). There were wild eustatic fluctuations that did cause changes in the shelf populations (in fact, throughout the Paleozoic), but nothing on the scale of the Permian extinctions. Finally, bolide-caused extinctions happen so much more swiftly, as the end-Cretaceous event shows, that the one well-documented case cannot be compared at all with what happened in the Permian. Paleo-Tethyan anoxia has a temporal and spatial evolution that best explains the paleontological and paleoenvironmental data at hand, especially in a world that was fairly rapidly becoming poorer in oxygen (Berner, 2004; Ward, 2006).[44]

But why the double peaks? What could have led Paleo-Tethys to erupt twice (if it really did so)? We know so little now that this would appear a premature question to most prudent minds. We are not even sure how sharp the double peaks really were. But if the double peaks persist in the years to come, a possible solution to such a problem readily presents itself within the limits of the presently available observations. What if the Paleo-Tethyan triple junction built a flood basalt province in the middle of the Paleo-Tethys, perhaps similar to the Mid-Pacific Mountains built during the earlier late Cretaceous near the Pacific-Farallon-Phoenix triple junction (Winterer, 1976)? A Guadalupian flood basalt province would be able to destabilize the accumulated gases both in the sediments and in the water column and lead to gas eruption. Unfortunately, attempts at dating and deciphering the petrology and geochemistry of the mafic tectonic lenses caught up within the Paleo-Tethyan accretionary complexes are so few that no fruitful discussion may be taken beyond this point on this subject, but it is a most promising avenue of attack to decipher the history of the Paleo-Tethyan waters.

---

[44] In his useful book *The First One Hundred Years of American Geology*, George Perkins Merrill (1924, p. 245–246) reports that Henry Darwin Rogers (together with his brother William Barton Rogers, of great Appalachian fame) read a paper to the Association of American Geologists and Naturalists meeting in Washington, D.C., in 1844 in which he expressed astonishment at the amount of coal produced in the late Carboniferous and speculated that this would have greatly augmented the relative abundance of oxygen in the atmosphere. He felt this "to be a matter of great interest in geology, as showing that every modification in the constitution of the air had adapted it to the development of animals progressively higher in the scale of organization. Singularly enough the paper seems to have excited little interest at the time" (Merrill, 1924, p. 246).

CHAPTER IX

Conclusions

All available evidence concerning the massive Permian extinctions, on land and at sea, converge to imply that their causative mechanism was the poisoning of the gases the organisms breathed in marine waters and in an atmosphere that were fast being reduced in oxygen content. Timing and regional geology show that the poisoning started in the abyssal depths of the closed, poorly aerated equatorial basin of the Paleo-Tethys already in the Carboniferous and spread gradually as the volume of poisoned water increased onto the shelves. Figure 27 illustrates this in a simplified manner, a sort of artist's conception of how all that might have happened.

The double-peaked Permian extinction may very well have been, in effect, a continuous event, but it could have left an episodic record because the biosphere apparently adapted itself to changing conditions in bouts of evolutionary change. After every change there was apparently a time of rapid extinction and slow adaptation, until the changing conditions again overwhelmed the new products of the evolution as so nicely seen in the diagram by Stanley and Yang (1994) showing the Permian adventures of the Chinese brachiopods (see our Fig. 23). One could in fact draw a very similar diagram for the tetrapods with small temporal offsets.

By the Triassic, the Paleo-Tethys had much diminished in size and by the end of the period it all but vanished. We know from vast areas in China and central Asia that many Paleo-Tethyan subduction-accretion complexes had surfaced by the Norian, adding large areas to Asia. In the west, rifting in the future Mediterranean region largely eliminated the closed geometry of the Paleo-Tethys. The once proud internal ocean of Pangea could no longer maintain its independence: the Ptolemaic condition largely disappeared and, with it, the anoxia that had proven so calamitous to life on at least half of the planet, which happened to be the densely populated part. The asylums of the Panthalassa repopulated the marine world while the asylums on the periphery of the Pangea repopulated the land. The ancestors of the mesophytic flora could now recolonize the lowlands where immense dinosaurs eventually devoured tons of vegetation per day.

This study brings to the fore an old problem in geology that had become its central concern with the work of Sir Charles Lyell. Lyell argued that global catastrophes of the kind Cuvier thought necessary to explain global biostratigraphy did not in fact happen. Lyell's objection had been foreshadowed by Alexander von Humboldt's criticism of Cuvier's methodology of biostratigraphic correlation. The great geographer pointed out that global fossil (we would now call it biostratigraphic zone) correlation was necessarily limited by the exigencies of global physical geography. Neither animals nor plants had universal distribution, but

were limited by climate and environment (de Humboldt, 1823, p. 52f.). Lyell added that nothing in the geological record showed the incidence of past *global* events. Contrary to the common perception of his views, Sir Charles was not against catastrophes. He in fact mentioned that a catastrophic emptying of the North American Great Lakes would cause continent-wide floods and such a thing he thought eminently plausible (Lyell, 1830, S. 89; he was almost fully vindicated by the shocking discovery, nearly a century later, of the catastrophic Spokane flood that had formed the Channeled Scablands in eastern Washington State [Bretz, 1923, 1925]).[45] But a flood covering the entire Earth had never happened and he could not think how it could happen if the present economy of nature also ruled the events in the past.

This sort of thinking soon led to what Suess called "quietism"[46] and it was thought incapable of explaining the success of the global stratigraphy. Darwin believed the incompleteness of the geological record was the answer, but Suess pointed out that some of the caesurae were far too widespread to fit comfortably into Darwin's scheme (Suess, 1904, p. 265), and the present consensus of opinion agrees with Suess's view (Jablonski, 2005, p. 193). Suess's solution to the dilemma was to assume local events that might have global repercussions. He thought, for example, that local oceanic subsidences would cause global regressions and thus global unconformities (see Şengör, 2006).

Suess also saw that the progress of life on Earth had not been steady and in fact that very unsteadiness was the basis of the great success of biostratigraphy. But the challenge was how to generate an unsteady, punctuated—to use a term that Eldredge and Gould (1972) invented—in fact, "Cuvierian," life history in a world developing irregularly yet uniformly, in a "Lyellian" way, so to say. Suess could not solve this problem and so far nobody has been able to. Many of the recent attempts to do so often have taken recourse to a Cuvierian global catastrophism and either dreamed of processes of which we have no experience (such as large meteorite or comet impacts) or imagined that what we do know might have acted in immensely greater intensities in the

---

[45]Few geologists now remember that James Hutton (1726–1797) anticipated the Messinian salinity crisis within his actualistic world view that he propounded in his classical 1788 paper: "Let us but suppose a rock placed across the gut of Gibraltar, (a case nowise unnatural), and the bottom of the Mediterranean would be certainly filled with salt, because the evaporation from the surface of that sea exceeds the measure of its supply" (Hutton, 1788, p. 242).

[46]"Quietism" is a word that had been invented by the Archbishop of Naples, Cardinal Innico Caracciolo (senior), in a letter he wrote to the Pope Innocent XI on 30 January 1682, to describe the movement started by the Spanish ascetic and Roman Catholic priest Miguel de Molinos (1640–1697). This movement exalted inactivity to allow God to have uninterfered sway over the human spirit. For a short history of this strange doctrine, see Paquier (1910). Suess thought it properly symbolized Sir Charles's view of geology.

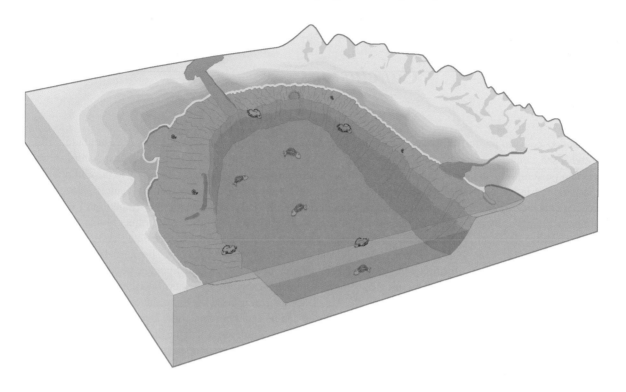

Carboniferous	Permian						Triassic	
	Cis		Gua		Lop		E. Triassic	

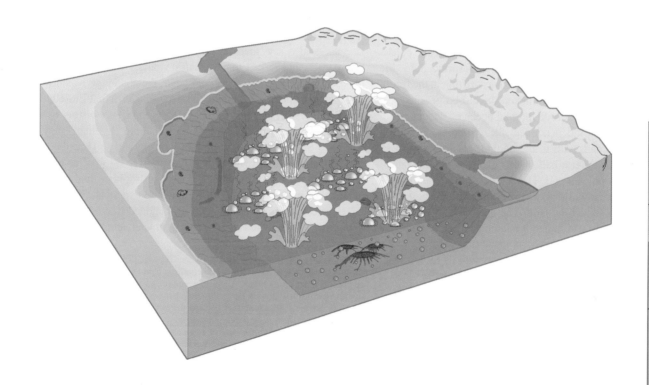

Carboniferous	Permian						Triassic	
	Cis		Gua		Lop		E. Triassic	

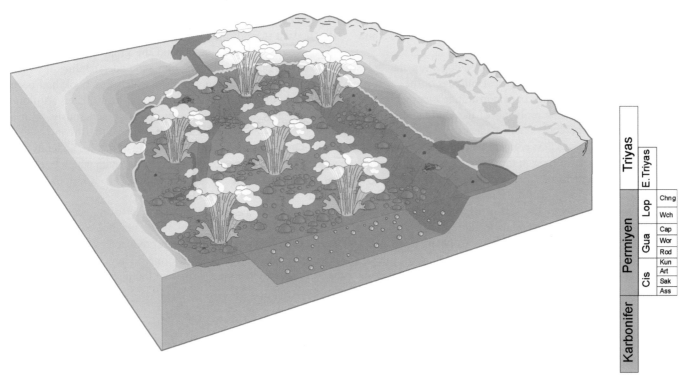

Figure 27 (*on this and previous page*). Artist's conception of what might have happened in the Paleo-Tethys during the Carboniferous to Triassic interval. (A) Medial Guadalupian: Abyssal anoxia in the Paleo-Tethys, which might have commenced already in the Carboniferous, has already reached the bathyal areas and is about to spill over into the Panthalassa via the northern Thailand strait(s). Abyssal benthic life is all dead. Necton largely escaped to asylums, possibly in the Panthalassa. (B) Late Guadalupian: Anoxia has already reached some of the lower shelves. Suboceanic igneous vulcanicity may have triggered a first phase of oceanic gas eruptions. (C) Changhsingian: Almost all of Paleo-Tethys has turned anoxic. Only some protected littoral asylums may have escaped it. Major gas eruptions may have resulted from oversaturation of the water column by dissolved gases because of millions of years of anoxia and/or further submarine igneous vulcanicity. Widespread death brings the Permian period to a close and with it the Paleozoic era. The Mesozoic begins with a much impoverished biosphere.

past (e.g., vast and rapid trap eruptions). Philosophically, neither of these two procedures is acceptable, as James Hutton and, following in his footsteps, Sir Charles so emphatically put forward. The past is gone and has left little record. What is incumbent on the student of historical geology is to try to fit, best as she or he can, what little record there is to the processes she or he is familiar with first hand, i.e., to the present changes taking place before our very eyes. Even what is going on now is not nearly as accessible to us as we would wish. But the *principle* ought to be to stick to what can be known for sure and then to try to extrapolate into the past using the fragmentary record of history as far as can be done. That is why the present world must be investigated in as much detail as possible and *all lines of evidence* from the past must be used in reconstructing it in terms of the present-day world. Not all our attempts to squeeze the past into the straightjacket of the present will be crowned with success, however. But then, we must erect *testable* hypotheses to explain, albeit temporarily, what we think the past might have been like. One problem that emerges here is the imprecision of all kinds of geological data. It is therefore important to check data sets from different processes influencing each other to see whether any incongruence may be

detected. A good knowledge of the present world imparts on the geologist a good intuition to be able to make sound judgments in such cases. An extremely critical requirement in formulating hypotheses is to state when we would consider a hypothesis of ours definitively falsified.

In discussions on the Permian extinctions a tendency has developed to try to explain it using certain individual methodologies. For instance, the volcanic hypothesis has so far put immense reliance on *dating* the trap eruptions, but its practitioners have been careless about the temporal and spatial aspects of the extinctions to be explained. The bolide impactors have only too hastily saw a repetition of the end-Cretaceous event in every extinction and have relied most heavily on *geochemical data*. Paleontologists were generally satisfied when they could explain extinctions among the organisms they knew best (and a surprisingly large number proved most enthusiastic to revert to a Cuvierian view of the world using such excuses as the Signor-Lipps effect). Stratigraphers claimed with confidence global synchronies by *biostratigraphy,* not rarely when none could really be demonstrated. However, unlike its service disciplines, physics and chemistry, or the mathematical tools it uses, natural history advances not by sticking to techniques, but to problems. Many techniques stemming from a number of service disciplines may bear on the solution to a problem.

If this study has any merit in shedding even the tiniest amount of light on one of the greatest enigmas of the earth and life sciences, it possibly lies in demonstrating that the methodology of approaching the problems of the kind tackled in this paper via regional geology in its broadest sense—regional stratigraphy, regional tectonics, regional paleontology—is still the most powerful tool we have in the natural sciences. To explain geological phenomena all other observations must be hung on the framework regional geology forms. There is little hope of tackling the problems of the sort of complexity the Permian extinctions represent by pretending to study "processes" alone. Study of processes is useful in helping us broaden our observations and reducing their results to relationships we can take in at a glance. Study of processes can only be done by studying their products preserved for us in our planet through the vicissitudes of natural history. Nothing more. Without honoring the vast store of geological and biological data, we cannot hope even to begin to understand the grand problems of our beloved planet, to the appreciation and understanding of which the laureate of the symposium, to which this book had been originally destined, Bill Dickinson, has contributed so gloriously and has taught to so many of us so generously, with a view to enlarging human understanding and to ushering us to a common joy before the grand spectacle of nature. So let us end with the immortal words of Eduard Suess, with which he closed his grand work on the geology of Earth, *The Face of the Earth*, and its last chapter entitled "Life":

"In face of these open questions let us rejoice in the sunshine, the starry firmament and all the manifold diversity of the Face of the Earth, which has been produced by these very processes, recognizing, at the same time, to how great a degree life is controlled by the peculiarities of the planet and its fortunes" (Suess, 1909, p. 777).

CHAPTER X

References Cited

Adamia, S., ed., 1984, Georgian Soviet Socialist Republic excursions, *in* International Geological Congress, 27th, Moscow: Tibilisi, Khelovneba Publishing House, 224 p.

Adamia, S., Belov, A.A., Lordkipanidze, M., and Somin, M.L., 1982, Correlation of Prevariscan and Variscan events in the Alpine Mediterranean Mountain Belt, *in* IGCP Field Excursion Guidebook of International Working Meeting on the Caucasus, Tbilisi: Paris, International Geoscience Program, project no. 15, 83 p.

Adams, A.S., and Weaver, C.E., 1958, Thorium-to-uranium ratios as indicators of sedimentary processes: example of concept of geochemical facies: AAPG Bulletin, v. 42, p. 387–430.

Adatte, T., 2003, Evolution de la biosphère et extinctions en masse, conjonctions des causes: exemple de la limite Crétacé-Tertiaire (KT): Bulletin de la Société Neuchâteloise des Sciences Naturelles, v. 126, p. 5–27.

Ager, D.V., 1971, Space and time in brachiopod history, *in* Middlemiss, F.A., Rawson, P.F., and Newall, G., eds., Faunal Provinces in Space and Time, Proceedings of the 17th Inter-University Geological Congress, Queen Mary College, University of London: Liverpool, Seel House Press, p. 95–110.

Akçiz, S., Burchfiel, B.C., Crowley, J.L., Yin, J.Y., and Chen, L.Z., 2008, Geometry, kinematics, and regional significance of the Chong Shan shear zone, Eastern Himalayan Syntaxis, Yunnan, China: Geosphere, v. 4, p. 292–314, doi: 10.1130/GES00111.1.

Algeo, T.J., Hannigan, R., Rowe, H., Brookfield, M., Baud, A., Krystyn, L., and Ellwood, B.B., 2007, Sequencing events across the Permian-Triassic boundary, Guryul Ravine (Kashmir, India): Palaeogeography, Palaeoclimatology, Palaeoecology, v. 252, p. 328–346, doi: 10.1016/j.palaeo.2006.11.050.

Ali, J.R., Thompson, G.M., Song, X., and Wang, Y., 2002, Emeishan basalts (SW China) and the "end-Guadalupian" crisis: magnetobiostratigraphic constraints: Journal of the Geological Society, v. 159, p. 21–29, doi: 10.1144/0016-764901086.

Ali, J.R., Thompson, G.M., Zhou, M.F., and Song, X.Y., 2005, Emeishan large igneous province, SW China: Lithos, v. 79, p. 475–489, doi: 10.1016/j.lithos.2004.09.013.

Allen, M.B., and Vincent, S.J., 1997, Fault reactivation in the Junggar region, northwest China: the role of basement structures during Mesozoic-Cenozoic compression: Journal of the Geological Society, v. 154, p. 151–155, doi: 10.1144/gsjgs.154.1.0151.

Allen, M.B., Şengör, A.M.C., and Natal'in, B.A., 1995, Junggar, Turfan, and Alakol basins as Late Permian to ?Early Triassic sinistral shear structures in the Altaid orogenic collage, Central Asia: Journal of the Geological Society, v. 152, p. 327–338, doi: 10.1144/gsjgs.152.2.0327.

Altıner, D., 1981, Recherches stratigraphiques et micropaléontologiques dans le Taurus oriental au NW de Pınarbaşı (Turquie) [Ph.D. thesis]: Université de Genève, Section des Sciences de la Terre, no. 2005, 450 p.

Altıner, D., and Özgül, N., 2001, Carboniferous and Permian of the Allochthonous Terranes of the central Tauride belt, southern Turkey, in PaleoForams 2001, International Conference on Paleozoic Benthic Foraminifera Guidebook, 20–24 August 2001: Ankara, Turkey, Middle East Technical University, 36 p.

Altıner, D., Özkan-Altıner, S., and Koçyiğit, A., 2000, Late Permian foraminiferal biofacies belts in Turkey: palaeogeographic and tectonic implications, *in* Bozkurt, E., Winchester, J.A., and Piper, J.D.A., eds., Tectonics and Magmatism in Turkey and the Surrounding Area: Geological Society (London) Special Publication 173, p. 83–96.

Alvarez, L., Alvarez, W., Asaro, F., and Michel, M., 1980, Extraterrestrial cause for the Cretaceous-Tertiary mass extinction: Science, v. 208, p. 1095–1108, doi: 10.1126/science.208.4448.1095.

Alvarez, W., 1997, *T. rex* and the Crater of Doom: Princeton, Princeton University Press, 185 p.

Alve, E., and Goldstein, S.T., 2003, Propagule transport as a key to method of dispersal in benthic foraminifera (Protista): Limnology and Oceanography, v. 48, p. 2163–2170.

Amalitzky, V., 1901, Sur la découverte, dans les dépôts permiens supérieurs du nord de la limite de la Russie, d'une flore glossoptérienne et de reptiles Pareiasaurus et Dicynodon: Comptes-rendus hébdomadaires des Séances de l'Académie de Sciences (Paris), v. 132, p. 591–593.

Andrusov, N.I., 1890, Predvaritel'nii atchet ob uchastii b Chernomorskoi glubomernoi ekspeditsii 1890 g.: Izvestiya Rossikoye Goegraficheskoye Obshestva, v. 26, vip. 2, geogr. izv., p. 398–409.

Angielczyk, K.D., and Walsh, M.L., 2008, Patterns in the evolution of nares size and secondary palate length in anomodont Therapsids (Synapsida): implications for hypoxia as a cause of end-Permian Tetrapod extinctions: Journal of Paleontology, v. 82, p. 528–542, doi: 10.1666/07-051.1.

Angiolini, L., Brunton, H., and Gaetani, M., 2005, Early Permian (Asselian) brachiopods from Karakorum (Pakistan) and their palaeobiogeographical significance: Palaeontology, v. 48, no. 1, p. 69–86, doi: 10.1111/j.1475-4983.2004.00439.x.

Anonymous, 1979, Stratigraficheskii Kodeks SSSR—Vremennii Svod Pravil i Rekomendatsii, Utverzhden Mezhdomstvennim Stratigraficheskim Komitetom CCCR 10 Maya 1976 g.: Leningrad, Ministerstvo Geologii SSSR Vsesoyuznii Ordena Lenina Nauchno-Issledovatel'skii Geologicheskii Institut (VSEGEI), Akademiya Nauk SSSR, Mezhdomstvennii Stratigraficheskii Komitet SSSR, 148 p.

Aplonov, S.V., 1989, The paleogeodynamics of the West Siberian Platform: International Geology Review, v. 31, p. 859–867.

Archibald, J.D., 1996, Dinosaur Extinction and the End of an Era: New York, Columbia University Press, Critical Moments in Paleobiology and Earth History Series, 237 p.

Argand, E., 1924, La tectonique de l'Asie, *in* Congrés Géologiques International, Comptes Rendus de la XIIIme session, Premier Fascicule, H. Vaillant-Carmanne, Liége, p. 171–372.

Avouac, J.P., Tapponnier, P., Bai, M., You, H., and Wang, G., 1993, Active thrusting and folding along the northern Tien Shan and late Cenozoic rotation of the Tarim relative to Dzungaria and Kazakhstan: Journal of Geophysical Research, v. 98, p. 6755–6804, doi: 10.1029/92JB01963.

Bambach, R.K., Knoll, A.H., and Wang, S.C., 2004, Origination, extinction and mass depletions of marine diversity: Paleobiology, v. 30, p. 522–542, doi: 10.1666/0094-8373(2004)030<0522:OEAMDO>2.0.CO;2.

Battail, B., 1997, Les genres *Dicynodon* et *Lystrosaurus* (Therapsida, Dicynodotia) en Eurasie: une mise au point: Geobios, v. 20, p. 39–48, doi: 10.1016/S0016-6995(97)80007-1.

Battail, B., 2000, A comparison of late Permian Gondwanan and Laurasian amniote faunas: Journal of African Earth Sciences, v. 31, p. 165–174, doi: 10.1016/S0899-5362(00)00081-6.

Battail, B., Dejax, J., Richir, P., Taquet, P., and Ve'ran, M., 1995, New data on the continental Upper Permian in the area of Luang-Prabang, Laos, *in* Proceedings of the IGCP Symposium on Geology of SE Asia, Hanoi, XI/1995, Geological Survey of Vietnam: Journal of Geology, series B, no. 5–6, p. 11–14.

Baturin, G.N., 1971, Stages of phosphorite formation on the sea floor: Nature Physical Science (London), v. 232, p. 61–62.

Baud, A., Richoz, S., and Marcoux, J., 2005, Calcimicrobial cap rocks for the basal Triassic units: western Taurus occurrences (SW Turkey): Comptes Rendus Palévol, v. 4, p. 569–582, doi: 10.1016/j.crpv.2005.03.001.

Baud, A., Marcoux, J., and Richoz, S., 2006, Oceanic record of the Permian-Triassic crisis: view from Tethys (Hawasina, Oman) and comparison with Panthalassa (Accretate terranes), *in* Lüer, V., Hollis, C., Campbell, H., and Simes, J., eds., Inter Rad 11 and Triassic Stratigraphy Symposium: a joint conference hosted by the International Association of Radiolarian paleontologists, IGCP 467 and the Subcommission on Triassic Stratigraphy: Wellington, New Zealand, Institute of Geological and Nuclear Sciences Limited, p. 10.

Beerling, D.J., and Berner, R.A., 2005, Feedbacks and the coevolution of plants and atmospheric CO_2: Proceedings of the National Academy of Sciences of the United States of America, v. 102, p. 1302–1305, doi: 10.1073/pnas.0408724102.

71

Bender, F.K., and Raza, H.K., 1995, Geology of Pakistan: Beiträge zur Regionalen Geologie der Erde, v. 25, Gebrüder Borntraeger, Berlin, 414 p. + 10 foldouts.

Ben Ferjani, A., Burollet, P.F., and Mejri, F., 1990, Petroleum Geology of Tunisia: Tunis, Entreprise Tunisienne d'Activités Pétrolières, 193 p. + 5 plates.

Benton, M.J., 2003, When Life Nearly Died—The Greatest Extinction of All Time: London, Thames & Hudson Ltd., 336 p.

Benton, M.J., 2004, Origin and relationships of Dinosauria, *in* Weishampel, D.B., Dodson, P., and Osmólska, H., eds., The Dinosauria, second edition: Berkeley, University of California Press, p. 7–19.

Berner, R.A., 1981, A new geochemical classification of sedimentary environments: Journal of Sedimentary Petrology, v. 51, p. 359–365.

Berner, R.A., 1999, Atmospheric oxygen through Phanerozoic time: Proceedings of the National Academy of Sciences of the United States of America, v. 96, p. 10,955–10,957, doi: 10.1073/pnas.96.20.10955.

Berner, R.A., 2003, The rise of trees and their effects on Paleozoic atmospheric CO_2 and O_2: Comptes Rendus Geoscience, v. 335, p. 1173–1177, doi: 10.1016/j.crte.2003.07.008.

Berner, R.A., 2004, The Phanerozoic Carbon Cycle: CO_2 and O_2: Oxford, Oxford University Press, 150 p.

Berner, R.A., 2005, The carbon and sulfur cycles and atmospheric oxygen from middle Permian to middle Triassic: Geochimica et Cosmochimica Acta, v. 69, p. 3211–3217, doi: 10.1016/j.gca.2005.03.021.

Berner, R.A., 2006, GEOCARBSULF: A combined model for Phanerozoic atmospheric O_2 and CO_2: Geochimica et Cosmochimica Acta, v. 70, p. 5653–5664, doi: 10.1016/j.gca.2005.11.032.

Berner, R.A., 2007, The effect of land plant evolution on Paleozoic atmospheric O_2: Geological Society of America Abstracts with Programs, v. 39, no. 6, p. 24.

Berner, R.A., Beerling, D.J., Dudley, R., Robinson, J.M., and Wildman, R.A., Jr., 2003, Phanerozoic atmospheric oxygen: Annual Review of Earth and Planetary Sciences, v. 31, p. 105–134, doi: 10.1146/annurev.earth.31.100901.141329.

Berner, R.A., VandenBrooks, J.M., and Ward, P.D., 2007, Oxygen and evolution: Science, v. 316, p. 557–558, doi: 10.1126/science.1140273.

Bernoulli, D., 1972, North Atlantic and Mediterranean Mesozoic facies: a comparison, *in* Hollister, C.D., and Ewing, J.L., et al., Deep Sea Drilling Project Initial Reports: Washington, D.C., Government Printing Office, v. 11, p. 801–871.

Bernoulli, D., 2007, The pre-Alpine geodynamic evolution of the Southern Alps: a short summary: Bulletin der angewandten Geologie, v. 12/2, p. 3–10.

Bernoulli, D., and Jenkyns, H.C., 1974, Alpine, Mediterranean and central Atlantic Mesozoic facies in relation to the early evolution of the Tethys, *in* Dott, R.H., Jr., and Shaver, R.H., eds., Modern and Ancient Geosynclinal Sedimentation: Society for Sedimenary Geology (SEPM) Special Publication 19, p. 129–160.

Bernoulli, D., and Laubscher, H., 1972, The palinspastic problem of the Hellenides: Eclogae Geologicae Helvetiae, v. 65, p. 107–118.

Berthelsen, A., 1998, The Tornquist Zone northwest of the Carpathians: an intraplate-pseudosuture: Geologiska Föreningens i Stockholm Förhandlingar, v. 120, p. 223–230.

Betz, D., Führer, F., Greiner, G., and Plein, E., 1987, Evolution of the Lower Saxony basin: Tectonophysics, v. 137, p. 127–170, doi: 10.1016/0040-1951(87)90319-2.

Beuerlen, K., 1962, Der Faunenschnitt an der Perm-Trias Grenze: Zeitschrift der Deutschen Geologischen Gesellschaft, v. 108, p. 88–99.

Beuerlen, K., 1975, Geologie—Die Geschichte der Erde und des Lebens: Kosmos Gesellschaft der Naturfreunde, Frank'sche Verlagshandlung, Stuttgart, 318 p.

Bhatti, N.A., 1977, Phosphorite deposits of the Kalkul-Mirpur area, Hazara District, N.W. Frontier Province, Pakistan, *in* Notholt, A.J.G., ed., Phosphate Rock in the CENTO Region (Iran, Pakistan and Turkey): Ankara, Turkey, Working Group on Phosphates, 125 p.

Bhaud, M., 1998, The spreading potential of polychaete larvae does not predict adult distributions; consequences for conditions of recruitment: Hydrobiologia, v. 375–376, p. 35–47, doi: 10.1023/A:1017073409259.

Bjerager, M., Seidler, L., Stemmerik, L., and Surlyk, F., 2006, Ammonoid stratigraphy and sedimentary evolution across the Permian-Triassic boundary in east Greenland: Geological Magazine, v. 143, p. 635–656, doi: 10.1017/S0016756806002020.

Blendinger, W., Van Vliet, A., and Hughes-Clarke, M.W., 1990, Updoming, rifting and continental margin development during the Late Paleozoic in northern Oman: Geological Society, London, Special Publication 49, p. 27–37, doi: 10.1144/GSL.SP.1992.049.01.03.

Bommeli, R., 1890, Die Geschichte der Erde: Stuttgart, J.B.W. Dietz, 684 p. + 2 plates.

Bosellini, A., 1964, Stratigrafia, petrografia e sedimentologia delle facies carbonatice al limite Permo-Trias nelle Dolomiti occidentali: Memorie del Museo di Storia Naturale della Venezia-Tridentino, v. 15, p. 59–110.

Bradley, D.C., 2007, Age distribution of passive margins through earth history and tectonic implications: Eos (Transactions of the American Geophysical Union), fall meeting supplement, abs. U44A-03.

Bradshaw, M.T., Yeats, A.N., Beynon, R.M., Brakel, A.T., Langford, R.P., Totterdell, J.M., and Yeung, M., 1988, Palaeogographic evolution of the northwest shelf, *in* Purcell, P.G., and Purcell, R.R., eds., The North West Shelf Australia—Proceedings North West Shelf Symposium: Perth, W.A., Petroleum Exploration Society of Australia, p. 29–54.

Bramlette, M.N., 1965, Massive extinction of biota at the end of Mesozoic time: Science, v. 148, p. 1696–1699, doi: 10.1126/science.148.3678.1696.

Brandner, R., 1987, Ozeanographische und klimatische Veränderungen an der Perm-Trias-Grenze in der westlichen Tethys: Heidelberger Geowissenschaftliche Abhandlungen, v. 8, p. 44–45.

Brasier, M.D., 1988, Foraminiferal extinction and ecological collapse during global biological events, *in* Larwood, G.P., ed., Extinction and Survival in the Fossil Record: Oxford, Systematics Association and Clarendon Press, p. 37–64.

Brayard, A., Escarguel, G., and Bucher, H., 2007, The biogeography of Early Triassic ammonoid faunas: clusters, gradients, and networks: Géobios, v. 40, p. 749–765, doi: 10.1016/j.geobios.2007.06.002.

Bretz, J.H., 1923, The Channeled Scabland of the Columbia Plateau: The Journal of Geology, v. 31, p. 617–649.

Bretz, J.H., 1925, The Spokane flood beyond the Channeled Scablands: The Journal of Geology, v. 33, p. 97–115, 236–259.

Broglio Loriga, C., and Cassinis, G., 1992, The Permo-Triassic boundary in the Southern Alps (Italy) and in adjacent Periadriatic regions, *in* Sweet, W.C., Yang, Z.Y., and Yin, H.F., eds., Permo-Triassic Events in the Eastern Tethys—Stratigraphy, Classfication and Relations with the Western Tethys: Cambridge, Cambridge University Press, p. 78–97.

Brookfield, M., Twitchett, R.J., and Goodings, C., 2003, Palaeoenvironments of the Permian-Triassic transition sections in Kashmir, India: Palaeogeography, Palaeoclimatology, Palaeoecology, v. 198, p. 353–371, doi: 10.1016/S0031-0182(03)00476-0.

Buffetaut, E., 1984, The palaeobiogeographical significance of the Mesozoic continental vertebrates from south-east Asia, *in* Buffetaut, E., Jaeger, J.-J., and Rage, J.-C., eds., Paléogéographie de l'Inde, du Tibet et du Sud-Est Asiatique: Confrontations des Données Paléontologiques aves les Modèles Géodynamiques: Mémoires de la Société Géologique de France, nouvelle série no. 147, p. 37–42.

Buffetaut, E., 1989, The contribution of vertebrate palaeontology to the geodynamic history of south east Asia, *in* Şengör, A.M.C., ed., The Tectonic Evolution of the Tethyan Region: Dordrecht, Kluwer Academic Publishers, p. 645–653.

Buffetaut, E., 2003, La Fin des Dinosaures—Comment les Grandes Extinctions ont Façonné le Monde Vivant: Paris, Fayard, 243 p.

Bunopas, S., and Vella, P., 1983, Tectonic and geologic evolution of Thailand, *in* Proceedings of a Workshop on Stratigraphic correlation of Thailand and Malaysia: Bangkok, Geological Society of Thailand, and Kuala Lumpur, Geological Society of Malaysia, v. 1, p. 307–322.

Burchfiel, B.C., Zhang, P.Z., Wang, Y.P., Zhang, W.Q., Song, F.M., Deng, Q.D., Molnar, P., and Royden, L.H., 1991, Geology of the Haiyuan fault zone, Ningxia-hui autonomous region, China, and its relation to the evolution of the northeastern margin of the Tibetan Plateau: Tectonics, v. 10, p. 1091–1110, doi: 10.1029/90TC02685.

Burchfiel, B.C., Cowan, D.S., and Davis, G.A., 1992a, Tectonic overview of the Cordilleran orogen in the western United States, *in* Burchfiel, B.C., Lipman, P.W., and Zoback, M.L., eds., The Cordilleran Orogen: conterminous U.S.: Boulder, Geological Society of America, Geology of North America, v. G-3, p. 407–479.

Burchfiel, B.C., Zhang, P.Z., Wang, Y.P., Zhang, W.Q., Song, F.M., Deng, Q.D., Molnar, P., and Royden, L.H., 1992b, Correction to "Geology of the Haiyuan fault zone, Ningxia-hui Autonomous Region, China, and its relation to the evolution of the northeastern margin of the Tibetan Plateau" by B.C. Burchfiel et al.: Tectonics, v. 11, p. 175, doi: 10.1029/92TC00001.

Bureau of Geology and Mineral Resources of Anhui Province, 1987, Regional Geology of Anhui Province: Beijing, Geological Publishing House, People's Republic of China, Ministry of Geology and Mineral Resources, Geological Memoirs, series 1, no. 5, 721 p.

Bureau of Geology and Mineral Resources of Sichuan Province, 1991, Regional Geology of Sichuan Province: Beijing, Geological Publishing House, People's Republic of China, Ministry of Geology and Mineral Resources, Geological Memoirs, series 1, no. 23, 730 p.

Burke, K., 1977, Aulacogens and continental breakup: Annual Review of Earth and Planetary Sciences, v. 5, p. 371–396, doi: 10.1146/annurev.ea.05.050177.002103.

Burke, K., and Şengör, A.M.C., 1988, Ten metre global sea-level change associated with south Atlantic Aptian salt deposition: Marine Geology, v. 83, p. 309–312, doi: 10.1016/0025-3227(88)90064-3.

Burnett, W.C., 1977, Geochemistry and origin of phosphorite deposits from off Peru and Chile: Geological Society of America Bulletin, v. 88, p. 813–823, doi: 10.1130/0016-7606(1977)88<813:GAOOPD>2.0.CO;2.

Burov, E.B., and Molnar, P., 1998, Gravity anomalies over the Fergana Valley (central Asia) and intracontinental deformation: Journal of Geophysical Research, v. 103, p. 18,137–18,152, doi: 10.1029/98JB01079.

Burrett, C., and Stait, B., 1986, China and Southeast Asia as part of the Tethyan margin of Cambro-Ordovician Gondwanaland, *in* McKenzie, K.G., ed., International Symposium on Shallow Tethys 2, Wagga-Wagga, 15–17 September 1986: Rotterdam, A.A. Balkema, p. 65–77.

Burtman, V.S., 1994, Meso-Tethyan oceanic sutures and their deformation: Tectonophysics, v. 234, p. 305–327, doi: 10.1016/0040-1951(94)90230-5.

Burtman, V.S., Skobelev, S.F., and Molnar, P., 1996, Late Cenozoic slip on the Talas-Ferghana fault, the Tien Shan, central Asia: Geological Society of America Bulletin, v. 108, p. 1004–1021, doi: 10.1130/0016-7606(1996)108<1004:LCSOTT>2.3.CO;2.

Buser, S., Grad, K., Ogorelec, B., Ramovš, A., and Šribar, L., 1986, Stratigraphical, paleontological and sedimentological characteristics of Upper Permian beds in Slovenia, NW Yugoslavia: Memorie della Societa Geologica Italiana, v. 34, p. 195–210.

Busson, G., 1969, Rapports entre terrains Mésozoïques et Paléozoïques au Sahara algéro-tunisien: la discordance hercynienne: Comptes Rendus hébdomadaires de l'Académie des Sciences de Paris, v. 269, p. 685–688.

Busson, G., and Cornée, A., 1996, L'événement océanique anoxique du Cénomanien supérieur-terminal: une revue et une interprétation mettant en jeu une stratification des eaux marines par le CO_2 mantellique: Société Géologique du Nord Publication, no. 23, 143 p.

Çağatay, A., Pehlivanoğlu, H., and Altun, Y., 1979–1980, Küre piritli bakır yataklarının kobalt-altın mineralleri ve bu metaller açısından ekonomik değeri: Maden Tetkik ve Arama Enstitüsü Dergisi, no. 93/94, p. 110–117.

Callomon, J.H., and Donovan, D.T., 1966, Stratigraphic classification and terminology: Geological Magazine, v. 103, p. 97–99.

Carey, S.W., 1958, The tectonic approach to continental drift, *in* Carey, S.W., convener, Continental Drift—A Symposium: Hobart, Geology Department, University of Tasmania, p. 177–358.

Carroll, R.L., 1988, Vertebrate Paleontology and Evolution: New York, W.H. Freeman and Company, 698 p.

Carulli, G.B., Radrizzani, C.P., and Ponton, M., 1986, The Permian-Triassic boundary in the Paularo area (Carnia): Memorie della Societa Geologica Italiana, v. 34, p. 107–120.

Casshyap, S.M., 1982, Palaeodrainage and palaeogeography of Son-Valley Gondwana basin, Madhya Pradesh, *in* Valdiya, K.S., Bhatia, S.B., and Gaur, V.K., eds., Geology of Vindhyānchal: Delhi, Hindustan Publishing Corporation (India), p. 132–142.

Cassinis, G., Neri, C., and Perotti, C.R., 1993, The Permian and Permian-Triassic boundary in eastern Lombardy and western Trentino (Southern Alps, Italy), *in* Lucas, S.G., and Morales, M., eds., The nonmarine Triassic: Albuquerque, New Mexico Museum of Natural History and Science Bulletin 3, p. 51–63.

Cassinis, G., Toutin-Morin, N., and Virgili, C., 1995, A general outline of the Permian continental basins in southwestern Europe, *in* Scholle, P.A., Peryt, T.M., and Ulmer-Scholle, D.S., eds., The Permian of Northern Pangea, Sedimentary Basins and Economic Resources: Berlin, Springer-Verlag, v. 2, p. 137–157.

Cassinis, G., Nicosia, U., Lozovsky, V.R., and Gubin, Y.M., 2002, A view on the Permian continental stratigraphy of the Southern Alps, and general correlation with the Permian of Russia: Permophiles, v. 40, p. 4–16.

Ceoloni, P., Conti, M.A., Mariotti, N., and Nicosia, U., 1988, New Late Permian tetrapod footprints from Southern Alps, *in* Cassinis, G., ed., SGI and IGCP Project 203, Proceedings of the Field Conference on "Permian and Permian-Triassic boundary in the South-Alpine segment of the Western Tethys, and additional regional reports," Brescia, 4–12 July 1986: Memorie della Società Geologica Italiana, v. 34 (1986), p. 45–65.

Chai, C., Zhou, Y., Mao, X., Ma, S., Ma, J., Kong, P., and He, J., 1992, Geochemical constraints on the Permo–Triassic boundary event in south China, *in* Sweet, W.C., Jr., Yang, Z., Dickins, J.M., and Yin, H., eds., Permo–Triassic Events in the Eastern Tethys: Cambridge, Cambridge University Press, p. 158–168.

Chaloner, W.G., and Creber, G.T., 1988, Fossil plants as indicators of late Palaeozoic plate positions, *in* Audley-Charles, M.G., and Hallam, A., eds., Gondwana and Tethys: Geological Society, London, Special Publication 37, p. 201–210.

Chaloner, W.G., and Meyen, S.V., 1973, Carboniferous and Permian floras of the northern continents, *in* Hallam, A., ed., Atlas of Palaeobiogeography: Amsterdam, Elsevier, p. 169–186.

Chamberlin, T.C., 1909, Diastrophism as the ultimate basis of correlation: The Journal of Geology, v. 17, p. 685–693.

Chang, C.F., Pan, Y.S., and Sun, Y.Y., 1989, The tectonic evolution of Qinghai-Xizang Plateau: a review, *in* Şengör, A.M.C., ed., The Tectonic Evolution of the Tethyan Region: Dordrecht, Kluwer Academic Publishers, p. 415–476.

Charig, A.J., 1984, Competition between therapsids and archosaurs during the Triassic period: a review and synthesis of current theories: Symposia of the Zoological Society of London, v. 52, p. 597–628.

Chen, H.H., Sun, S., Li, J.L., Hsü, K.J., Heller, F., and Dobson, J., 1992, Preliminary paleomagnetic evidence for early Triassic separation of Huanan and Yangtze blocks, south China: Chinese Science Bulletin, v. 137, p. 1297–1301.

Chen, H.H., Sun, S., Li, J.L., Heller, F., Dobson, J., Haag, M., and Hsü, K.J., 1993, Early Triassic paleomagnetism and tectonics, south China: Journal of Southeast Asian Earth Sciences, v. 8, p. 269–276, doi: 10.1016/0743-9547(93)90028-N.

Chen, Y., 1992, Evolution tectonique le long d'une Transversale entre Inde et Siberie—Apports de nouvelles données paléomagnétiques crétacées du Tibet, du Tarim et de la Dzungarie [Ph.D. thèse]: Paris, l'Université de Paris VII, 306 p.

Chen, Y., and Courtillot, V., 1989, Widespread Cenozoic (?) remagnetization in Thailand and its implications for the India–Asia collision: Earth and Planetary Science Letters, v. 93, p. 113–122.

Chen, Y., Cogné, J.-P., and Courtillot, V., 1992, New Cretaceous paleomagnetic poles from the Tarim Basin, nothwestern China: Earth and Planetary Science Letters, v. 114, p. 17–38, doi: 10.1016/0012-821X(92)90149-P.

Chen, Y., Gilder, S., Halim, N., Cogné, J.P., and Courtillot, V., 2002, New paleomagnetic constraints on central Asian kinematics: Displacement along the Altyn Tagh fault and rotation of the Qaidam Basin: Tectonics, v. 21, no. 5, p. 1042, doi: 10.1029/2001TC901030.

Choubert, B., 1935, Recherches sur la genèse des chaînes paléozoïques et anté-cambriennes: Revue de Géographie Physique et de Géologie Dynamique, v. 8, p. 5–50.

Cita, M.B., 2006, Exhumation of Messinian evaporites in the deep-sea and creation of deep anoxic brine-filled collapsed basins: Sedimentary Geology, v. 188–189, p. 357–378, doi: 10.1016/j.sedgeo.2006.03.013.

Cobbold, P.R., Sadybakasov, E., and Thomas, J.C., 1996, Cenozoic transpression and basin development, Kyrgyz Tienshan, central Asia, *in* Roure, F., Ellouz, N., Shein, V.S., and Skvortsov, I., eds., Geodynamic Evolution of Sedimentary Basins: Paris, Editions Technip, p. 181–202.

Colbert, E.H., 1975, Early Triassic tetrapods and Gondwanaland, *in* XVII Congrès International de Zoologie Monaco, 25–30 Septembre 1972 Thème N° 1 Biogéographie et Liaisons Intercontinentales au Cours du Mésozoïque: Mémoires du Muséum National d'Histoire Naturelle, nouvelle série, Série A, Zoologie, v. 88, p. 202–215.

Collin, P.-Y., Kershaw, S., Crasquin-Soleau, S., and Feng, Q.L., 2008, Facies changes and diagenetic processes across the Permian-Triassic boundary event horizon, Great Bank of Guizhou, South-China: a controversy of erosion and dissolution: Sedimentology, doi: 10.1111/j.1365-3091.2008.00992.x.

Conti, M.A., Leonardi, G., Mariotti, N., and Nicosia, V., 1977, Tetrapod Footprints of the Val Gardena Sandstone (North Italy): Their Paleontological, Stratigraphical and Paleoenvironmental Meaning: Paleontographica Italica, v. 70, 91 p.

Conti, M.A., Mariotti, N., Manni, R., and Nicosia, U., 2000, Tetrapod footprints in the Southern Alps: an overview, *in* Cassinis, G., Cortesogno, L., Gaggero, L., Massari, F., Neri, C., Nicosia, U., and Pittau, P., coordinators, Stratigraphy and Facies of the Permian Deposits Between Eastern Lombardy and the Western Dolomites, Field Trip Guidebook, second edition, The Continental Permian International Congress, 15–25 September 1999: Brescia, Italy, The Continental Permian International Congress, p. 137–138.

Coogan, A.H., 1960, Stratigraphy and paleontology of the Permian Nosoni and Dekkas Formations (Bollibokka Group): University of California Publications in Geological Sciences, v. 36, p. 243–316, plates 22–27.

Cosgriff, J.W., 1965, A new genus of Temnospondyli from the Triassic of western Austrialia: Journal and Proceedings of the Royal Society of Western Australia, v. 48, p. 65–90.

Counillon, H., 1896, Documents pour servir à l'étude géologique des environs de Luang-Prabang (Cochinchine): Comptes Rendus hébdomadaire de l'Academie des Sciences (Paris), v. 123, p. 1330–1333.

Courtillot, V., 1999, Evolutionary Catastrophes—The Science of Mass Extinction: Cambridge, Cambridge University Press, 173 p.

Courtillot, V., Jaupart, C., Manighetti, I., Tapponnier, P., and Besse, J., 1999, On causal links between flood basalts and continental breakup: Earth and Planetary Science Letters, v. 166, p. 177–195, doi: 10.1016/S0012-821X(98)00282-9.

Crasquin, S., Perri, M.C., Nicora, A., and de Wever, P., 2008, Ostracods across the Permian-Triassic boundary in western Tethys: the Bulla stratotype (Southern Alps, Italy): Rivista Italiana di Paleontologiia e Stratigrafia, v. 114, p. 233–262.

Crasquin-Soleau, S., and Kershaw, S., 2005, Ostracod fauna from the Permian-Triassic boundary interval of South China (Huaying Mountains, eastern Sichuan Province): palaeoenvironmental significance: Palaeogeography, Palaeoclimatology, Palaeoecology, v. 217, p. 131, doi: 10.1016/j.palaeo.2004.11.027.

Crasquin-Soleau, S., Marcoux, J., Angiolini, L., and Nicora, A., 2004, Palaeocopida (Ostracoda) across the Permian-Triassic events: new data from southwestern Taurus (Turkey): Journal of Micropalaeontology, v. 23, no. 1, p. 67–76.

Crowley, T.J., 1994, Pangean climates, in Klein, G.D., ed., Pangea: Paleoclimate, Tectonics and Sedimentation During Accretion, Zenith, and Breakup of a Supercontinent: Boulder, Geological Society of America Special Paper 288, p. 25–39.

Cunningham, D., Owen, L.A., Snee, L.W., and Li, J.L., 2003, Structural framework of a major intracontinental orogenic termination zone: the easternmost Tien Shan, China: Journal of the Geological Society, v. 160, p. 575–590, doi: 10.1144/0016-764902-122.

Cunningham, W.D., Windley, B.F., Dorjnamjaa, D., Badamgarov, J., and Saandar, M., 1996, Late Cenozoic transpression in southwestern Mongolia and the Gobi Altai-Tien Shan connection: Earth and Planetary Science Letters, v. 140, p. 67–81, doi: 10.1016/0012-821X(96)00048-9.

Cuvier, G., 1796, Mémoire sur les espèces d'Eléphans tant vivantes que fossiles, lu à la séance publique de l'Institut national le 15 germinal, an IV: Magasin Encyclopédique, 2. année, no. 3, p. 440–445.

Cuvier, G., 1812, Recherches sur les Ossemens Fossiles de Quadrupeds où l'on rétablit les caractères de plusieurs espèces d'animaux que les révolutions du globe paroissent avoir détruites: Deterville, Paris, v. 1, not consecutively paginated.

Cuvier, G., and Geoffroy [Saint-Hilaire, E.], 1791–1799 (1795 April/June issue), Sur les espèces d'Eléphans, par CC. Cuvier et Geoffroy: Bulletin des Sciences de la Société Philomatique de Paris, série 1, v. 1, p. 90.

Dacqué, E., 1936, Aus der Urgeschichte der Erde und des Lebens—Tatsachen und Gedanken: München, R. Oldenbourg, 230 p.

Dadlez, R., 2006, The Polish basin—relationship between the crystalline, consolidated and sedimentary crust: Geological Quarterly, v. 50, p. 43–58.

Damiani, R., Neveling, J., Hancox, J., and Rubidge, B., 2000, First trematosaurid temnospondyl from the Lystrosaurus Assemblage Zone of South Africa and its biostratigraphic implications: Geological Magazine, v. 137, p. 659–665, doi: 10.1017/S0016756800004660.

Dana, J.D., 1847a, Geological results of the earth's contraction in consequence of cooling: American Journal of Science and Arts, 2nd series, v. 3, p. 176–188.

Dana, J.D., 1847b, Origin of the continents: American Journal of Science and Arts, 2nd series, v. 3, p. 94–100.

Dana, J.D., 1863, Manual of Geology: Treating the Principles of the Science with Special Reference to American Geological History: Philadelphia, Theodore Bliss & Co., 798 p. + 1 map.

Darby, B.J., Davis, G.A., and Zheng, Y.D., 2001, Structural evolution of the southwestern Daqing Shan, Yinshan Belt, Inner Mongolia, in Hendrix, M.S., and Davis, G.A., eds., Paleozoic and Mesozoic Tectonic Evolution of Central and East Asia: From Continental Assembly to Intracontinental Deformation: Boulder, Geological Society of America Memoir 194, p. 199–214.

Darby, B.J., Davis, G.A., Zhang, X.H., Wu, F.Y., Wilde, S., and Yang, J.H., 2004, The newly-discovered Waziyu metamorphic core complex, Yiwulü Shan, western Liaoning Province: northwest China: Earth Science Frontiers, v. 11, p. 145–155.

Davis, G.A., 2003, The Yanshan belt of north China: tectonics, adakitic magmatism, and crustal evolution: Earth Science Frontiers, v. 10, p. 373–384.

Davis, G.A., Zheng, Y.D., Wang, C., Darby, B.J., Zhang, C.H., and Gehrels, G., 2001, Mesozoic tectonic evolution of the Yanshan fold and thrust belt, with emphasis on Hebei and Liaoning provinces, northern China, in Hendrix, M.S., and Davis, G.A., eds., Paleozoic and Mesozoic Tectonic Evolution of Central and Eastern Asia: From Continental Assembly to Intracontinental Deformation: Boulder, Geological Society of America Memoir 194, p. 171–197.

Davis, G.A., Darby, B., Zheng, Y.D., and Spell, T.L., 2002, Geometric and temporal evolution of an extensional detachment fault, Hohhot metamorphic core complex, Inner Mongolia, China: Geology, v. 30, p. 1003–1006, doi: 10.1130/0091-7613(2002)030<1003:GATEOA>2.0.CO;2.

Davis, G.A., Xu, B., Zheng, Y.D., and Zhang, W.J., 2004, Indosinian extension in the Solonker suture zone: the Sonid Zuoqi metamorphic core complex, Inner Mongolia, China: Earth Science Frontiers, v. 11, p. 135–144.

de Cserna, Z., 1960, Orogenesis in time and space in Mexico: Geologische Rundschau, v. 50, p. 595–605, doi: 10.1007/BF01786872.

de Humboldt, A., 1823, Essai Géognostique sur le Gisement des Roches dans les Deux Hémisphères: Paris, F.G. Levrault, 379 p.

Delville, N., Arnaud, N., Montel, J.-M., Roger, F., Brunel, M., Tapponnier, P., and Sobel, E., 2001, Paleozoic to Cenozoic deformation along the Altyn Tagh Fault in the Altun Shan massif area, eastern Qilian Shan, northeastern Tibet, China, in Hendrix, M.S., and Davis, G.A., eds., Paleozoic and Mesozoic Tectonic Evolution of Central and Eastern Asia: From Continental Assembly to Intracontinental Deformation: Boulder, Geological Society of America Memoir 194, p. 269–292.

Desio, A., 1963, Review of the geologic "formations" of the western Karakorum (central Asia): Rivista Italiana di Paleontologia e Stratigrafia, v. 69, p. 475–501.

Desmond, A.J., 1977, The Hot-Blooded Dinosaurs—A Revolution in Palaeontology: New York, Warner Books, 352 p.

De Wever, P.D., O'Dogherty, L., and Goričan, S., 2006, The plankton turnover at the Permo-Triassic boundary, emphasis on radiolarians: Eclogae Geologicae Helvetiae, v. 99, supplement 1, p. 49–62, doi: 10.1007/s00015-006-0609-y.

Dewey, J.F., Pitman, W.C., III, Ryan, W.B.F., and Bonnin, J., 1973, Plate tectonics and the evolution of the Alpine system: Geological Society of America Bulletin, v. 84, p. 3137–3180, doi: 10.1130/0016-7606(1973)84<3137:PTATEO>2.0.CO;2.

Dewey, J.F., Hempton, M.R., Kidd, W.S.F., Şaroğlu, F., and Şengör, A.M.C., 1986, Shortening of continental lithosphere: the neotectonics of Eastern Anatolia—a young collision zone, in Coward, M.P., and Ries, A.C., eds., Collision Tectonics: Geological Society, London, Special Publication 19 (R.M. Shackleton volume), p. 3–36.

Dickinson, W.R., 1992, Cordilleran sedimentary assemblages, in Burchfiel, B.C., Lipman, P.W., and Zoback, M.L., eds., The Cordilleran Orogen: conterminous U.S.: Boulder, Geological Society of America, Geology of North America, v. G-3, p. 539–551.

Dickinson, W.R., 2004, Evolution of the North American Cordillera: Annual Review of Earth and Planetary Sciences, v. 32, p. 13–45, doi: 10.1146/annurev.earth.32.101802.120257.

Dicks, D.R., 1960, The Geographical Fragments of Hipparchus: London, University of London, The Athlone Press, 214 p.

Diener, C., 1895, Ergebnisse einer Geologischen Expedition in den Central-Himalaya von Johar, Hundes, und Painkhanda: Denkschriften der kaiserlichen Akademie der Wissenschaften (Wien), mathematisch-naturwissenschaftliche Klasse, v. 62, p. 533–606 + 7 plates.

Dimichele, W.A., Phillips, T.L., and Peppers, R.A., 1985, The influence of climate and depositional environment on the distribution of Pennsylvanian coal swamps, in Tiffney, B.H., ed., Geological Factors and the Evolution of Plants: New Haven, Yale University Press, p. 223–256.

Ding, M.H., 1986, Permian-Triassic boundary and conodonts in south China: Memorie della Società Geologica Italiana, v. 34, p. 263–268.

Dolenec, M., 2005, The Permian/Triassic boundary in the Karavanke Mountains (Brsnina section, Slovenia): The ratio of Th/U as a possible indicator of changing redox conditions at the P/T transition: RMZ Materials and Geoenvironment, v. 52, p. 437–445.

Dolenec, T., Lojen, S., Buser, S., and Dolenec, M., 1999, Stable isotope event markers near the Permo-Triassic boundary in the Karavanke Mountains (Slovenia): Geologia Croatica, v. 52, p. 77–81.

Donovan, S.K., ed., 1989, Mass Extinctions: Processes and Evidence: Ferdinand Enke, Stuttgart, 266 p.

D'Orbigny, A., 1852, Cours Élémentaire de Paléontologie et de Géologie Stratigraphiques, tome second: Paris, Victor Masson, 848 p.

Drooger, C.W., 1974, The boundaries and limits of stratigraphy: Koninklijke Nederlandse Akademie van Wetenschappen—Amsterdam, Proceedings, series B, v. 77, no. 3, p. 159–176.

Dubertret, L., ed., 1963, Lexique Stratigraphique International, Asia, Fascicule 1 (République Populaire Chinoise, I): Paris, Centre National de la Recherche Scientifique, 743 p.

Dumitru, T.A., and Hendrix, M.S., 2001, Fission-track constraints on Jurassic folding and thrusting in southern Mongolia and their relationship to the Beishan thrust belt of northern China, in Hendrix, M.S., and Davis, G.A., eds., Paleozoic and Mesozoic Tectonic Evolution of Central and Eastern Asia: From Continental Assembly to Intracontinental Deformation: Boulder, Geological Society of America Memoir 194, p. 215–229.

Eldredge, N., and Gould, S.J., 1972, Punctuated equilibria: An alternative to phyletic gradualism, in Schopf, T.J.M., ed., Models in Paleobiology: San Francisco, Freeman, Cooper and Co., p. 82–115.

Élie de Beaumont, L., 1852, Notice sur les Systèmes de Montagnes, volume three: Paris, P. Bertrand, p. 1069–1543 + 4 plates.

Ellis, R., 2004, No Turning Back—The Life and Death of Animal Species: New York, HarperCollins Publishers, 428 p.

Enkin, R., Yang, Z.Y., Yan, C., and Courtillot, V., 1992, Paleomagnetic constraints on the geodynamic history of the major blocks of China from the Permian to the Present: Journal of Geophysical Research, v. 97, p. 13,953–14,022, doi: 10.1029/92JB00648.

Erinç, S., 1984, Ortam Ekolojisi ve Degradasyonal Ekosistem Değişiklikleri: İstanbul Üniversitesi Deniz Bilimleri ve Coğrafya Enstitüsü Yayınları No: 1, İstanbul Üniversitesi Yayınları, no. 3213, 144 p.

Ernst, W.G., Tsujimori, T., Zhang, R., and Liou, J.G., 2007, Permo-Triassic collision, subduction zone metamorphism, and tectonic exhumation along the east Asian continental margin: Annual Review of Earth and Planetary Sciences, v. 35, p. 73–110, doi: 10.1146/annurev.earth.35.031306.140146.

Erwin, D.H., 1993, The Great Paleozoic Crisis—Life and Death in the Permian: New York, Columbia University Press, Critical Moments in Paleobiology and Earth History Series, 327 p.

Erwin, D.H., 2006, Extinction—How Life on Earth Nearly Ended 250 Million Years Ago: Princeton, Princeton University Press, 296 p.

Fan, C.J., 2000, Regional geology of the Sichuan-Yunnan-Qinghai-Xizang domain, in Chen, Y.Q., ed., Concise Regional Geology of China: Beijing, Geological Publishing House, p. 132–171.

Farabegoli, E., Perri, C., and Posenato, R., 2007, Environmental and biotic changes across the Permian-Triassic boundary in western Tethys: The Bulla stratotype, Italy: Global and Planetary Change, v. 55, p. 109–135, doi: 10.1016/j.gloplacha.2006.06.009.

Fastovsky, D.E., and Smith, J.B., 2004, Dinosaur paleoecology, in Weishampel, D.B., Dodson, P., and Osmólska, H., eds., The Dinosauria, second edition: Berkeley, University of California Press, p. 614–626.

Fischer, J., 1932a, Clavdii Ptolemaei Geographiae Codex Vrbinas Graecvs 82 Phototypice Depictvs Consilio et Opera Cvratorvm Bibliothecae Vaticanae: Codices e Vaticanis Selecti, vol. XVIIII, E.J. Brill, Leiden and Otto Harrasowitz, Leipzig, Tomvs Prodromvs (de Cl.Ptolemaei Vita Operibvs Geographia Praesertim Eisvsque Fatis Pars Rior Commentatio: 605 p.), Pars Prior Textvs (facsimile of manuscript) cvm appendice critica Pii Franchi de Cavalieri (37 p. + 76 plates), Part Altera (Tabvlae Geographicae LXXXIII Graecae - Arabicae - Latinae e Codicivs LIII Selectae).

Fischer, J., 1932b [1991], Introduction, in Geography of Claudius Ptolemy: New York, New York Public Library, p. 3–15 (reprinted in 1991 by Dover Publications).

Flesch, L.M., Haines, A.J., and Holt, W.E., 2001, Dynamics of the India-Eurasia collision zone: Journal of Geophysical Research, v. 106, p. 16,435–16,460, doi: 10.1029/2001JB000208.

Flügel, H.W., and Hubmann, B., 1993, Paläontologie und Plattentektonik am Beispiel proto- und paläotethyder Korallenfaunen: Jahrbuch der Geologischen Bundesanstalt, v. 136, p. 27–37.

Foote, M., 2003, Origination and extinction through the Phanerozoic: a new approach: The Journal of Geology, v. 111, p. 125–148, doi: 10.1086/345841.

Franke, W., 2000, The mid-European segment of the Variscides: tectonostratigraphic units, terrane boundaries and plate tectonic evolution, in Franke, W., Haak, V., Oncken, O., and Tanner, D., eds., Orogenic Processes: Quantification and Modelling in the Variscan Belt: Geological Society, London, Special Publication 179, p. 35–61.

Franquin, I., Riboulleau, A., Bodineau, L., Thoumelin, G., Tribovillard, N., and Tannenbaum, E., 2006, Préservation de matière organique et conditions de depôt des phosphorites d'Israël (Crétace supérieur): Les matières Organiques en France—État de l'Art et Prospectives 22–24 Janvier 2006 Carqueiranne, Résumés, p. I-14.

Gaetani, M., 1997a, The Karakorum block in central Asia, from Ordovician to Cretaceous: Sedimentary Geology, v. 109, p. 339–359, doi: 10.1016/S0037-0738(96)00068-1.

Gaetani, M., 1997b, The north Karakorum in the framework of the Cimmerian blocks: Himalayan Geology, v. 18, p. 33–47.

Gaetani, M., undated [2006], Geologia del Nord Karakorum—Stato dell'Arte: no publisher, no place of publication [Milano], 24 p.

Gaetani, M., Anigolini, L., Garzanti, E., Jadoul, F., Leven, E.Y., Nicora, A., and Sciunnach, D., 1995, Permian stratigraphy in the northern Karakorum, Pakistan: Rivista Italiana di Paleontologia e Stratigrafia, v. 101, p. 107–152 + 11 plates.

Gaetani, M., Le Fort, P., Tanoli, S., Angiolini, L., Nicora, A., Sciunnach, D., and Kahn, A., 1996, Reconnaisance geology in Upper Chitral, Baroghil and Karambar districts (northern Karakorum, Pakistan): Geologische Rundschau, v. 85, p. 683–704, doi: 10.1007/BF02440105.

Gaetani, M., Burg, J.P., Zanchi, A., and Jan, Q.M., 2004, A geological transect from the Indian plate to the east Hindu Kush, Pakistan, in 32nd International Geological Congress, Florence, Italy, August 20–28, 2004, Prestige Field Trip PR01: Firenze, Italy, International Geological Congress, 56 p.

Gansser, A., 1964, The Geology of the Himalayas: Malden, Massachusetts, Interscience Publishers, John Wiley & Sons, 289 p. + 4 plates.

Gebauer, E.V.I., 2007, Phylogeny and evolution of the Gorgonopsia with special reference to the skull and skeleton of GPIT/RE/7113 ('Aelurognathus?' parringtoni) [Ph.D. thesis]: Tübingen, Eberhard Karl University, 316 p.

Geluk, M.C., 2000, Late Permian (Zechstein) carbonate-facies maps, the Netherlands: Geologie & Mijnbouw, v. 79, p. 17–27.

Gilder, S.A., Gomez, J., Chen, Y., and Cogné, J.-P., 2008, A new paleogeographic configuration of the Eurasian landmass resolves a paleomagnetic paradox of the Tarim Basin (China): Tectonics, v. 27, TC1012, doi: 10.1029/2007TC002155.

Glennie, K., Higham, J., and Stemmerik, L., 2003, Permian, in Evans, D., ed., The Millenium Atlas: Petroleum Geology of the Central and Northern North Sea: London, The Millenium Atlas Company Limited, p. 91–103.

Gobbett, D.J., 1973, Permian fusulinacea, in Hallam, A., ed., Atlas of Palaeobiogeography: Amsterdam, Elsevier, p. 151–157.

Goddard, E.N., Trask, P.D., Deford, R.K., Rove, O.N., Singlewald, J.T., Jr., and Overbeck, R.M., 1975, Rock-Color Chart: Boulder, Geological Society of America.

Golonka, J., 2002, Plate tectonic maps of the Phanerozoic, in Kiessling, W., Flügel, E., and Golonka, J., eds., Phanerozoic Reef Patterns: SEPM Special Publication 72, p. 21–75.

Goodrich, E.S., 1916, On the classification of the Reptilia: Proceedings of the Royal Society of London, v. 89B, p. 261–276.

Gore, A., 2006, An Inconvenient Truth: London, Bloomsbury, 325 p.

Görür, N., and Şengör, A.M.C., 1992, The palaeogeographic and tectonic evolution of the eastern Tethysides: implications for the NW Australian margin breakup history, in von Rad, U., and Haq, B.U., et al., Proceedings of the Ocean Drilling Program, scientific results, v. 122: College Station, Texas, Ocean Drilling Program, p. 83–106.

Gradstein, F., Ogg, J., and Smith, A.G., eds., 2004, A Geologic Time Scale: Cambridge, Cambridge University Press, 588 p.

Graham, J.B., Dudley, R., Aguilar, N.M., and Gans, C., 1995, Implications of the late Palaeozoic oxygen pulse for physiology and evolution: Nature, v. 375, p. 117–120, doi: 10.1038/375117a0.

Grasmück, K., and Trümpy, R., 1969, Triassic stratigraphy and general geology of the country around Fleming Fjord (East Greenland), in Defretin-Lefranc, S., Grasmück, K., and Trümpy, R., eds., Notes on Triassic Stratigraphy and Paleontology of North-Eastern Jameson Land (East Greenland), Meddelelser Om Grønland, v. 168, no. 2, p. 5–71 + 4 plates.

Greene, T.J., Carroll, A.R., Hendrix, M.S., Graham, S.A., Wartes, M.A., and Abbink, O.A., 2001, Sedimentary record of Mesozoic deformation and inception of the Turpan-Hami basin, northwest China, in Hendrix, M.S., and Davis, G.A., eds., Paleozoic and Mesozoic Tectonic Evolution of Central and Eastern Asia: From Continental Assembly to Intracontinental Deformation: Boulder, Geological Society of America Memoir 194, p. 317–340.

Gregory, J.T., 1955, Vertebrates in the geologic time scale, *in* Poldervaart, A., ed., Crust of the Earth (A Symposium): Geological Society of America Special Paper 62, p. 593–608.

Gregory, J.W., 1915, Suess' classification of Eurasian mountains: The Geographical Journal for June 1915, p. 497–513.

Groves, J.R., and Altiner, D., 2005, Survival and recovery of calcareous foraminifera pursuant to the end-Permian mass extinction: Comptes Rendus Palévol, v. 4, p. 487–500, doi: 10.1016/j.crpv.2004.12.007.

Groves, J.R., Altiner, D., and Rettori, R., 2005, Extinction, survival, and recovery of Lagenide foraminifers in Permian-Triassic boundary interval, central Taurides, Turkey: Journal of Paleontology, v. 79, supplement to The Paleontological Society Memoir 4, v. 62, p. 1–38.

Güner, M., 1979–1980, Küre civarının masif sülfit yatakları ve jeolojisi, Pontidler (Kuzey Türkiye): Maden Tetkik ve Arama Enstitüsü Dergisi, no. 93/94, p. 65–109.

Gutnic, M., Monod, O., Poisson, A., and Dumont, J.-F., 1979, Géologie des Taurides Occidentales (Turquie): Mémoires de la Société Géologique de France, nouvelle série v. 58, 112 p.

Haas, J., Demény, A., Hips, K., Zajzon, N., Weiszburg, T.G., Sudar, M., and Pálfy, J., 2007, Biotic and environmental changes in the Permian-Triassic boundary interval recorded on a western Tethyan ramp in the Bükk Mountains, Hungary: Global and Planetary Change, v. 55, p. 136–154.

Hallam, A., 1969, Faunal realms and facies in the Jurassic: Palaeontology, v. 12, p. 1–18.

Hallam, A., 1987, End-Cretaceous mass extinction event: argument for terrestrial causation: Science, v. 238, p. 1237–1242, doi: 10.1126/science.238.4831.1237.

Hallam, A., 2004, Catastrophes and Lesser Calamities—The Causes of Mass Extinctions: Oxford, Oxford University Press, 226 p.

Hallam, A., and Wignall, P.B., 1997, Mass Extinctions and Their Aftermath: Oxford, Oxford University Press, 230 p.

Haug, E., 1900, Les Géosynclinaux et les aïres continentales. Contribution à l'étude des transgressions et regressions marines: Bulletin de la Société Géologique de France, 3 série, t. 28, p. 617–711.

Haug, E., 1907, Traité de Géologie, t. 1 (Les Phénomènes géologiques): Librairie Armand Colin, Paris, 538 p.

Haug, É., 1908–1911, Traité de Géologie, t. 2 (Les Périodes Géologiques): Librairie Armand Colin, Paris, p. 539–2024 + 64 plates.

Hayden, H.H., 1915, Notes on the geology of Chitral, Gilgit and the Pamirs: Records of the Gelogical Survey of India, v. 45, p. 271–320.

Hendrix, M.S., Graham, S.A., Carroll, A.R., Sobel, E.R., McKnight, C.L., Schulein, B.J., and Wang, Z.X., 1992, Sedimentary record and climatic implications of recurrent deformation in the Tien Shan: Evidence from Mesozoic strata of the north Tarim, Junggar, and Turpan basins, northwest China: Geological Society of America Bulletin, v. 104, p. 53–79, doi: 10.1130/0016-7606(1992)104<0053:SRACIO>2.3.CO;2.

Hendrix, M.S., Beck, M.A., Badarch, G., and Graham, S.A., 2001, Triassic synorogenic sedimentation in southern Mongolia: early effects of intracontinental deformation, *in* Hendrix, M.S., and Davis, G.A., eds., Paleozoic and Mesozoic Tectonic Evolution of Central and Eastern Asia: From Continental Assembly to Intracontinental Deformation: Boulder, Geological Society of America Memoir 194, p. 389–412.

Henningsen, D., and Katzung, G., 2002, Einführung in die Geologie Deutschlands, sixth edition: Berlin, Spektrum Akademischer Verlag Heidelberg, 214 p. + 16 plates.

Herak, M., 1999, Tectonic interrelation of the Dinarides and the Southern Alps: Geologia Croatica, v. 52, p. 83–98.

Heydari, E., Arzani, N., and Hassanzadeh, J., 2008, Mantle plume: The invisible serial killer—Application to the Permian-Triassic boundary mass extinctions: Palaeogeography, Palaeoclimatology, Palaeoecology, v. 264, p. 147–162, doi: 10.1016/j.palaeo.2008.04.013.

Hips, K., and Haas, J., 2006, Calcimicrobial stromatolites at the Permian-Triassic boundary in a western Tethyan section, Bükk Mountains, Hungary: Sedimentary Geology, v. 185, p. 239–253, doi: 10.1016/j.sedgeo.2005.12.016.

Holland, C.H., 1989, Trueman's epibole: Proceedings of the Geologists' Association, v. 100, p. 457–460.

Holt, W.E., Chamot-Rooke, N., Le Pichon, X., Haines, A.J., Shen-Tu, B., and Ren, J., 2000, Velocity field in Asia inferred from Quaternary fault slip rates and Global Positioning System observations: Journal of Geophysical Research, v. 105, p. 19,185–19,209, doi: 10.1029/2000JB900045.

Honegger, K., Dietrich, V., Frank, W., Gansser, A., Thöni, M., and Trommsdorff, V., 1982, Magmatism and metamorphism in the Ladakh Himalayas (the Indus-Tsangpo suture zone): Earth and Planetary Science Letters, v. 60, p. 253–292, doi: 10.1016/0012-821X(82)90007-3.

Hsü, J., 1976, On the discovery of a Glossopteris flora in southern Xizang and its significance in geology and palaeogeography: Scientica Geologica Sinica, v. 4, p. 324–331.

Hsü, K.J., 1971, Origin of the Alps and western Mediterranean: Nature, v. 233, p. 44–48, doi: 10.1038/233044a0.

Hsü, K.J., 1980, Terrestrial catastrophe caused by cometary impact at the end of Cretaceous: Nature, v. 285, p. 201–203, doi: 10.1038/285201a0.

Hsü, K.J., 1986, The Great Dying: San Diego, Harcourt Brace Jovanovich, 292 p.

Hsü, K.J., 1987, Catastrophic extinctions and the inevitability of the improbable: Dokumentation Nr. 28, Studiengruppe Energieperspektiven, p. 1–14.

Hsü, K.J., Sun, S., Chen, H.H., Pen, H.P., and Şengör, A.M.C., 1988, Mesozoic overthrust tectonics in south China: Geology, v. 16, p. 418–421, doi: 10.1130/0091-7613(1988)016<0418:MOTISC>2.3.CO;2.

Hsü, K.J., Li, J.L., Chen, H.H., Wang, Q.C., and Sun, S., 1990, Tectonics of south China: Key to understanding of west Pacific geology: Tectonophysics, v. 183, p. 9–39, doi: 10.1016/0040-1951(90)90186-C.

Hsu, R., Rigby, J., and Duan, S.Y., 1990, Revision of *Glossopteris* flora from southern Xizang: Scientia Geologica Sinica, no. 3, p. 233–242.

Huey, R.B., and Ward, P.D., 2005, Hypoxia, global warming and terrestrial late Permian extinctions: Science, v. 308, p. 398–401, doi: 10.1126/science.1108019.

Hutchison, C.S., 1989, Geological Evolution of South-east Asia: Oxford, Clarendon Press, Oxford Monographs on Geology and Geophysics, no. 13, 368 p.

Hutton, J., 1788, Theory of the Earth; or an investigation of the laws observable in the composition, dissolution, and restoration of land upon the globe: Transactions of the Royal Society of Edinburgh, v. 1, p. 209–304.

Isozaki, Y., 1997, Permo-Triassic boundary superanoxia and stratified superocean: records from lost deep sea: Science, v. 276, p. 235–238, doi: 10.1126/science.276.5310.235.

Isozaki, Y., 2006, Guadalupian (Middle Permian) giant bivalve Alatoconchidae from a mid-Panthalassan paleo-atoll complex in Kyushu, Japan: A unique community associated with Tethyan fusulines and corals: Proceedings of the Japanese Academy, series B, v. 82, p. 25–32.

Isozaki, Y., Yao, J.X., Matsuda, T., Sakai, H., Ji, Z.S., Simizu, N., Kobayashi, N., Kawahata, H., Nishi, H., Takano, M., and Kubo, T., 2004, Stratigraphy of the Middle-Upper Permian and lowermost Triassic at Chaotian, Sichuan, China: Proceedings of the Japanese Academy, v. 80, series B, p. 10–16.

Isozaki, Y., Kawahata, H., and Minoshima, K., 2007, The Capitanian (Permian) Kamura cooling event: the beginning of the Paleozoic-Mesozoic transition: Palaeoworld, v. 16, p. 16–30, doi: 10.1016/j.palwor.2007.05.011.

Jablonski, D., 2005, Mass extinctions and macroevolution: Paleobiology, supplement to v. 31 (Stephen Jay Gould volume, edited by Vrba, E.S., and Eldredge, N.), p. 192–210.

Jarvis, I., Burnett, W.C., Nathan, Y., Almbaydin, F., Attia, K.M., Castro, L.N., Flicoteaux, R., Hilmy, M.E., Hussain, V., Qutawna, A.A., Serjani, A., and Zanin, Y.N., 1994, Phosphorite Geochemistry: State of the Art and Environmental Concerns: Eclogae Geologicae Helvetiae, v. 87, p. 643–700.

Jenny, J.G., 1977, Géologie et stratigraphie de l'Elbourz Oriental entre Aliabad et Shahrud, Iran, thèse présentée à la Faculté des Sciences de l'Université de Genève pour obtenir le grade de docteur ès sciences, mention sciences de la terre, thèse no. 1820: Genève, Imprimérie Nationale, 238 p. + 4 plates + 1 appendix.

Jenny, J., and Stampfli, G., 1978, Lithostratigraphie du Permien de l'Elbourz oriental: Eclogae Geologicae Helvetiae, v. 71, p. 551–580.

Jin, Y., Wardlaw, B.R., Glenister, B.E., and Kotlyar, G.V., 1997, Permian chronostratigraphic subdivisions: Episodes, v. 20, p. 1015.

Johnson, K.S., Amsden, T.W., Denison, R.E., Dutton, S.P., Goldstein, A.G., Rascoe, B., Jr., Sutherland, P.K., and Thompson, D.M., 1988, Southern Midcontinent Region, *in* Sloss, L.L., ed., Sedimentary Cover—North American Craton: U.S.: Boulder, Geological Society of America, Geology of North America, v. D-2, p. 307–359.

Kahler, F., 1974, Fusuliniden aus T'ien-schan und Tibet Mit Gedanken zur Geschichte der Fusuliniden-Meere im Perm, *in* V. Invertebrate Palaeontology: Stockholm, The Sven Hedin Foundation, The Sino-Swedish Expedition Publication 52, v. 4, 148 p. + 2 plates.

Kajiwara, Y., Yamakita, S., Ishida, K., Ishida, H., and Imai, A., 1994, Development of a largely anoxic stratified ocean and its temporary massive mixing at the Permian/Triassic boundary supported by the sulfur isotopic record: Palaeogeography, Palaeoclimatology, Palaeoecology, v. 111, p. 367–379, doi: 10.1016/0031-0182(94)90072-8.

Kakuwa, Y., and Matsumoto, R., 2006, Cerium negative anomaly just before the Permian and Triassic boundary event—The upward expansion of anoxia in the water column: Palaeogeography, Palaeoclimatology, Palaeoecology, v. 229, p. 335–344, doi: 10.1016/j.palaeo.2005.07.005.

Kammer, T.W., Brett, C.E., Boardman, D.R., and Mapes, R.H., 1986, Ecologic stability of the dysaerobic biofacies during the late Paleozoic: Lethaia, v. 19, p. 109–121, doi: 10.1111/j.1502-3931.1986.tb00720.x.

Kato, Y., Nakao, K., and Isozaki, Y., 2002, Geochemistry of late Permian to early Triassic pelagic cherts from southwest Japan: implications for an oceanic redox change: Chemical Geology, v. 182, p. 15–34, doi: 10.1016/S0009-2541(01)00273-X.

Kauffman, E.G., and Erwin, D.H., 1995, Surviving mass extinctions: Geotimes, v. 40, no. 3, p. 14–17.

Kauffman, E.G., and Walliser, O.H., 1988, Global bioevents: abrupt changes in the global biota: Episodes, v. 11, p. 289–292.

Kemp, T.S., 2006, The origin of mammalian endothermy: a paradigm for the evolution of complex biological structure: Zoological Journal of the Linnean Society, v. 147, p. 473–488, doi: 10.1111/j.1096-3642.2006.00226.x.

Keppie, J.D., 2004, Mexican terranes revisited: A 1.3 Ga odyssey: International Geology Review, v. 46, p. 765–794, doi: 10.2747/0020-6814.46.9.765.

Khudoley, A.K., and Guriev, G.A., 1994, The formation and development of a late Paleozoic sedimentary basin on the passive margin of the Siberian paleocontinent, in Embry, A.F., Beauchamp, B., and Glass, D.J., eds., Pangea: Global Environmentsand Resources: Alberta, Canadian Society of Petroleum Geologists Memoir 17, p. 131–143.

Kiehl, J.T., and Shields, C.A., 2005, Climate simulation of the latest Permian: implication for mass extinction: Geology, v. 33, p. 757–760, doi: 10.1130/G21654.1.

King, G.M., 1988, Anomodontia, in Wellnhofer, P., ed., Handbuch der Paläoherpethologie, part 17C: Stuttgart, Gustav Fischer Verlag, 174 p.

King, G.M., and Jenkins, I., 1997, The dicynodont Lystrosaurus from the Upper Permian of Zambia: evolutionary and stratigraphical implications: Palaeontology, v. 40, p. 149–156.

Klaus, W., and Pak, E., 1974, Neue Beiträge zur Datierung von Evaporiten des Ober-Perm: Carinthia, v. 164, no. 84, p. 79–85.

Kling, G.W., 1987, Seasonal mixing and catastrophic degassing in tropical lakes, Cameroon, west Africa: Science, v. 237, p. 1022–1024, doi: 10.1126/science.237.4818.1022.

Kober, L., 1921, Der Bau der Erde: Berlin, Gebrüder Borntraeger, 324 p.

Kober, L., 1928, Der Bau der Erde, zweite neubearbeitete und vermehrte Auflage: Berlin, Gebrüder Borntraeger, 500 p. + 2 plates.

Kollmann, H.A., 1992, Tethys—the evolution of an idea: Schriftenreihe der Erdwissenschaftlichen Kommissionen der Österreichischen Akademie der Wissenschaften, v. 9, p. 9–14.

Kolodny, Y., 1981, Phosphorites, in Emiliani, C., ed., The Sea, volume 7: The Oceanic Lithosphere: New York, Wiley Interscience, John Wiley & Sons, p. 981–1023.

Kozur, H., 1977, Die Faunenänderungen nahe der Perm/Trias- und Trias/Jura-Grenze und ihre möglichen Ursachen, Teil I: Die Lage der Perm/Trias Grenze und die Änderung der faunen und Floren im Perm/Trias-Grenzbereich: Freiberger Forschungshefte, v. C326, p. 73–86.

Kozur, H., 1980, Die Faunenänderungen nahe der Perm=Trias und Trias=Jura-Grenze und ihre möglichen Ursachen. Teil II: Die Faunenänderungen an der Basis und innerhalb des Rhäts und die möglichen Ursachen für die Faunenänderungen nahe der Perm=Trias- und Trias=Jura-Grenze: Freiberger Forschungshefte, v. C357, p. 111–134.

Kozur, H., and Mostler, H., 1991, Pelagic Permian conodonts from an oceanic sequence at Sang-e Sefid (Fariman, NE-Iran): Abhandlunen der Geologischen Bundesanstalt in Wien, v. 38, p. 101–110.

Kozur, H.W., 1998, Some aspects of the Permian-Triassic boundary (PTB) and of the possible causes for the biotic crisis around this boundary: Palaeogeography, Palaeoclimatology, Palaeoecology, v. 143, p. 227–272, doi: 10.1016/S0031-0182(98)00113-8.

Kravchinsky, V.A., Cogné, J.-P., Herbert, W.P., and Kuzmin, M.I., 2002a, Evolution of the Mongol-Okhotsk Ocean as constrained by new palaeomagnetic data from the Mongol-Okhosk suture zone, Siberia: Geophysical Journal International, v. 148, p. 34–57, doi: 10.1046/j.1365-246x.2002.01557.x.

Kravchinsky, V.A., Konstantinov, K.M., Courtillot, V., Valet, J.-P., Savrasov, J.I., Cherniy, S.D., Mishenin, S.G., and Parasotka, B.S., 2002b, Palaeomagnetism of East Siberian traps and kimberlites: two new poles and palaeogeographic reconstructions at about 360 and 250 Ma: Geophysical Journal International, v. 148, p. 1–33, doi: 10.1046/j.0956-540x.2001.01548.x.

Królikowski, C., 2006, Crustal-scale complexity of the contact zone between the Palaeozoic platform and the east European craton in the NW Poland: Geological Quarterly, v. 50, p. 33–42.

Krull, E.S., Retallack, G.J., Campbell, H.J., and Lyon, G.L., 2000, $\delta^{13}C_{org}$ chemostratigraphy of the Permian-Triassic boundary in the Maitai Group, New Zealand: evidence for high-latitudinal methane release: New Zealand Journal of Geology and Geophysics, v. 43, p. 21–32.

Kump, L.R., Pavlov, A., and Arthur, M.A., 2005, Massive release of hydrogen sulfide to the surface ocean and atmosphere during intervals of oceanic anoxia: Geology, v. 33, p. 397–400, doi: 10.1130/G21295.1.

Labandeira, C., 2005, The fossil record of insect extinction: New approaches and future directions: American Entomologist, v. 51, p. 14–29.

Labandeira, C., and Sepkoski, J.J., Jr., 1993, Insect diversity in the fossil record: Science, v. 261, p. 310–315, doi: 10.1126/science.11536548.

Lacey, W.S., 1975, Some problems of 'mixed' floras in the Permian of Gondwanaland, in Gondwana Geology—Papers Presented at the Third Gondwana Symposium Canberra, Australia, 1973: Canberra, Australian National University Press, p. 125–134.

Laskarev, V., 1924, Sur les equivalents du Sarmatien supérieur en Serbie, in Vujevic, P., ed., Receuil des Travaux offert à M. Jovan Cvijič par ses Amis et Collaborateurs: Beograd, Drzhavna Shtamparija, p. 73–85.

Laubscher, H., and Bernoulli, D., 1977, Mediterranean and Tethys, in Nairn, A.E.M., Kanes, W.H., and Stehli, F.G., eds., The Ocean Basins and Margins, The Eastern Mediterranean: New York, Plenum Press, v. 4A, p. 1–28.

Le Conte, J., 1895, Critical periods in the history of the Earth: University of California, Bulletin of the Department of Geology, v. 1, p. 313–336.

Lefèvre, R., 1967, Un nouvel élément dans la géologie du Taurus lycien: les Nappes d'Antalya (Turquie): Comptes rendus hébdomadaire de l'Académie des Sciences (Paris), v. 265, p. 1365–1368.

Lepvrier, C., Maluski, H., Van Vuong, N., Roques, D., Axente, V., and Rangin, C., 1997, Indosinian NW-trending shear zones within the Truong Son belt (Vietnam) ^{40}Ar-^{39}Ar Triassic ages and Cretaceous to Cenozoic overprints: Tectonophysics, v. 283, p. 105–127, doi: 10.1016/S0040-1951(97)00151-0.

Lee, J.S., 1928, Some characteristic structural types in eastern Asia and their bearing upon the problem of continental movements: Geological Magazine, v. 66, p. 358–375, 413–431, 457–473, 501–522.

Less, G., Kovács, S., Pelikán, P., Pentelényi, L., and Sásdi, L., 2005, Geology of the Bükk Mountains, in Pelikán, P., ed., Regional Map Series of Hungary, Explanatory Book to the Geological Map of the Bükk Mountains (1:50,000): Budapest, Geological Institute of Hungary, 251 p. + 32 plates + colored map.

Leven, E.J., 1997, Permian Stratigraphy and Fusulinida of Afghanistan with Their Paleogeographic and Paleotectonic Implications: Geological Society of America Special Paper 316, 134 p.

Li, X.X., 1983, Note on three new species of Glossopteris flora from Qubu Formation, southern Xizang (Tibet), with discussion of the age of the formation: Acta Palaeontologica Sinica, v. 22, p. 130–138.

Littke, R., Bayer, U., Gajewski, D., and Nelskamp, S., eds., 2008, Dynamics of Complex Intracontinental Basins—The Central European Basin System: Berlin, Springer, 519 p. + CD-ROM.

Liu, H.Y., Liu, Y.J., and Hao, J., 1996, On the Banxi group and its related tectonic problems in south China: Journal of Southeast Asian Earth Sciences, v. 13, p. 191–196, doi: 10.1016/0743-9547(96)00025-6.

Liu, J.L., Davis, G.A., Lin, Z.Y., and Wu, F.Y., 2005, The Liaonan metamorphic core complex, southeastern Liaoning Province, north China: a likely contributor to Cretaceous rotation of eastern Liaoning, Korea and contiguous areas: Tectonophysics, v. 407, p. 65–80, doi: 10.1016/j.tecto.2005.07.001.

López-Gómez, J., Arche, A., and Pérez-López, A., 2002, Permian and Triassic, in Gibbons, W., and Moreno, T., eds., The Geology of Spain: London, The Geological Society, p. 185–212.

Lu, H.F., Howell, D.G., Jia, D., Cai, D.S., Wu, S.M., Chen, C.M., Shi, Y.S., Alin, Z.C., and Guo, L., 1994, Kalpin transpression tectonics, northwestern Tarim basin, western China: International Geology Review, v. 36, p. 975–981.

Lucas, S.G., 1998, Global Triassic tetrapod biostratigraphy and biochronology: Palaeogeography, Palaeoclimatology, Palaeoecology, v. 143, p. 347–384, doi: 10.1016/S0031-0182(98)00117-5.

Lucas, S.G., 2004, A global hiatus in the middle Permian tetrapod fossil record: Stratigraphy, v. 1, p. 47–64.

Lucas, S.G., 2006, Global Permian tetrapod biostratigraphy and biochronology, in Lucas, S.G., Cassinis, G., Schneider, J.W., eds., Non-Marine Permian Biostratigraphy and Biochronology: Geological Society, London, Special Publication 265, p. 65–93.

Lucas, S.G., and Heckert, A.B., 2001, Olson's gap: A global hiatus in the record of Middle Permian tetrapods: Journal of Vertebrate Paleontology, v. 21, p. 75A.

Lyell, C., (Sir), 1830, Principles of Geology, Being an Attempt to Explain the Former Changes of the Earth's Surface by Reference to Causes Now in Operation, volume 1: London, John Murray, 511 p.

Lyell, C., (Sir), 1833, Principles of Geology, Being an Attempt to Explain the Former Changes of the Earth's Surface by Reference to Causes Now in Operation, volume 3: London, John Murray, 398 p. + 4 plates + 1 map.

MacRae, C., 1999, Life Etched in Stone—Fossils of South Africa: Johannesburg, The Geological Society of South Africa, 305 p.

Magaritz, M., Bär, R., Baud, A., and Holser, W.T., 1988, The carbon-isotope shift at the Permian/Triassic boundary in the southern Alps is gradual: Nature, v. 331, p. 337–339, doi: 10.1038/331337a0.

Mahowald, N.M., Engelstaedter, S., Luo, C., Sealy, A., Artaxo, P., Benitez-Nelson, C., Bonnet, S., Chen, Y., Chuang, P.Y., Cohen, D.D., Dulac, F., Herut, B., Johansen, A.M., Kubilay, N., Losno, R., Maenhaut, W., Paytan, A., Prospero, J.M., Shank, L.M., and Siefert, R.L., 2009, Atmospheric iron deposition: global distribution, variability, and human perturbations: Annual Review of Marine Science, v. 1, p. 245–278, doi: 10.1146/annurev.marine.010908.163727.

Małkowski, K., 1982, Development and stratigraphy of the Kapp Starostin Formation (Permian) of Spitsbergen: Palaeontologia Polonica, v. 43, p. 69–81.

Mangerud, G., and Konieczny, R.M., 1993, Palynology of the Permian succession of Spitsbergen, Svalbard: Polar Research, v. 12, p. 65–93, doi: 10.1111/j.1751-8369.1993.tb00423.x.

Marcoux, J., 1987, Histoire et Topologie de la Néo-Téthys—Contribution à partir des exemples de la Turquie et de l'Himalaya-Tibet [doctorat d'état]: Paris, l'Université Pierre et Marie Curie, v. 1 (Introduction Générale, 73 + 10 p.), v. 2 (Sélection des Publications, 569 p.)

Martinez Garcia, E., 1990, Stephanian and Permian basins, in Dallmeyer, R.D., and Martinez Garcia, E., eds., Pre-Mesozoic Geology of Iberia: Berlin, Springer-Verlag, p. 39–54.

Martinson, D.G., and Pitman, W.C., III, 2007, The Arctic as a trigger for glacial terminations: Climatic Change, v. 80, p. 253–263, doi: 10.1007/s10584-006-9118-2.

Masaitis, V.L., 1983, Triassic and Permian volcanism of Siberia; problem of dynamic reconstruction: Mineralogicheskogo Obshchestva, v. 112, no. 4, p. 412–425.

Massari, F., 1986, Some thoughts on the Permo-Triassic evolution of the South-Alpine area (Italy): Memorie della Societa Geologica Italiana, v. 34, p. 179–188.

Mauritsch, H.J., and Becke, M., 1984, A magnetostratigraphic profile in the Permian (Gröden Beds, Val Gardena Formation) of the Southern Alps near Paularo (Carnic Alps, Friuli, Italy): IGCP No. 5 Newsletter, v. 5, p. 80–86.

McBride, E.F., 1974, Significance of color in red, green, purple, olive, brown, and grey beds of Difunta Group, northeastern Mexico: Journal of Sedimentary Petrology, v. 44, p. 760–773.

McGhee, G.R., Jr., 1996, The Late Devonian Mass Extinction—The Frasnian/Famennian Crisis: New York, Columbia University Press, 303 p.

McKerrow, W.S., and Ziegler, A.M., 1972, Palaeozoic oceans: Nature, v. 240, p. 92–94.

McLaughlin, J.C., 1980, Synapsida—A New Look into the Origins of Mammals: New York, The Viking Press, 148 p.

McQuarrie, N., Stock, J.M., Verdel, C., and Wernicke, B., 2003, Cenozoic evolution of Neotethys and implications for the causes of plate motions: Geophysical Research letters, v. 30, p. SDE 6-1–SDE 6-4, doi: 10.1029/2003GL017992.

Menning, M., Alekseev, A.S., Chuvashov, B.I., Davydov, V.I., Devuyst, F.X., Forke, H.C., Grunt, T.A., Hance, L., Heckel, P.H., Izokh, N.G., Jin, Y.G., Kotlyar, G.V., Kozur, H.W., Nemyrovskaya, T.I., Schneider, J.W., Wang, X.D., Weddige, K., Weyer, D., and Work, D.M., 2006, Global time scale and regional stratigraphic reference scales of central and west Europe, east Europe, Tethys, south China, and North America as used in the Devonian-Carboniferous-Permian Correlation Chart 2003 (DCP 2003): Palaeogeography, Palaeoclimatology, Palaeoecology, v. 240, p. 318–372.

Merla, G., 1930, La Fauna del Calcare a Bellrophon della Regione Dolomitica: Memoria dell'Istituto di Geologia Reale dell'Università di Padova, v. 9, 221 p.

Merrill, G.P., 1924, The First One Hundred Years of American Geology: New Haven, Yale University Press, 773 p. + 1 foldout plate.

Metcalfe, I., 1984, Stratigraphy, palaeontology and palaeogeography of the Carboniferous of Southeast Asia: Mémoire de la Société Géologique de France, no. 147, p. 107–118.

Metcalfe, I., 1996, Pre-Cretaceous evolution of SE Asian terranes, in Hall, R., and Blundell, D., eds., Tectonic Evolution of Southeast Asia: Geological Society, London, Special Publication 106, p. 97–122.

Métivier, F., and Gaudemer, Y., 1997, Mass transfer between eastern Tien Shan and adjacent basins (central Asia): constraints on regional tectonics and topography: Geophysical Journal International, v. 128, p. 1–17, doi: 10.1111/j.1365-246X.1997.tb04068.x.

Meyen, S.V., 1970, On the origin and relationship of the main Carboniferous and Permian floras and their bearing on general paleogeography of this period of time, in Second Gondwana Symposium South Africa July to August 1970 Proceedings and Papers: Pretoria, Council for Scientific and Industrial Research, Scientia, p. 551–555.

Meunier, S., 1917, Histoire Géologique de la Mer: Paris, Bibliothèque de Philosophie Scientifique, Ernst Flammarion, 324 p.

Michel, F., 2008, Le Tour de France d'un Géologue—Nos paysages Ont une Histoire: Paris, Delachaux et Niestlé, BRGM Éditions, 381 p.

Mihalynuk, M.G., 1997, Geology and mineral resources of the Tagish lake area, (NTS 104M/8,9,10E, 15 and 104N/12W), northwestern British Columbia: British Columbia Geological Survey Bulletin 105, 202 p. + 1 map.

Milanovskiy, Y.Y., 1976, Rift zones of the geologic past and their associated formations: International Geology Review, v. 18, p. 619–639.

Miller, E.L., Miller, M.M., Stevens, C.H., Wright, J.E., and Madrid, R., 1992, Late Paleozoic paleogeographic and tectonic evolution of the western U.S. Cordillera, in Burchfiel, B.C., Lipman, P.W., and Zoback, M.L., eds., The Cordilleran Orogen: conterminous U.S.: Boulder, Geological Society of America, Geology of North America, v. G-3, p. 57–106.

Mitchell, A.H.G., 1989, The Shan plateau and western Burma: Mesozoic-Cenozoic plate boundaries and correlations with Tibet, in Şengör, A.M.C., ed., The Tectonic Evolution of the Tethyan Region: Dordrecht, Kluwer Academic Publishers, p. 567–584.

Molnar, P., and Tapponnier, P., 1975, Cenozoic tectonics of Asia: Effects of a continental collision: Science, v. 189, p. 419–426, doi: 10.1126/science.189.4201.419.

Moore, J.C., 1975, Selective subduction: Geology, v. 3, p. 530–532, doi: 10.1130/0091-7613(1975)3<530:SS>2.0.CO;2.

Moseley, H.N., 1897 [1944], Notes by a Naturalist—An Account of Observations Made During the Voyage of H.M.S. "Challenger" Round the World in the Years 1872–1876: London, T. Werner Laurie, 540 p. + 1 map.

Murray, J.W., Stewart, K., Kassakian, S., Krynytzky, M., and DiJulio, D., 2007, Oxic, suboxic and anoxic conditions in the Black Sea, in Yanko-Hombach, V., Gilbert, A.S., Panin, N., and Dolukhanov, P.M., eds., The Black Sea Flood Question—Changes in Coastline, Climate and Human Settlement: Berlin, Springer, p. 1–21.

Musashi, M., Isozaki, Y., Koike, T., and Kreulen, R., 2001, Stable carbon isotope signature in mid-Panthalassa shallow-water carbonates across the Permo-Triassic boundary: evidence for ^{13}C-depleted superocean: Earth and Planetary Science Letters, v. 191, p. 9–20, doi: 10.1016/S0012-821X(01)00398-3.

Nakamura, K., Kimura, G., and Winsnes, T.S., 1987, Brachiopod zonation and age of the Permian Kapp Starostin Formation (Central Spitsbergen): Polar Research, v. 5, p. 207–219, doi: 10.1111/j.1751-8369.1987.tb00623.x.

Nakrem, H.A., 1994, Environmental distribution of bryozoans in the Permian of Spitsbergen, in Hayward, P.J., Ryland, J.S., and Taylor, P.D., eds., Biology and Palaeobiology of Bryozoans: Fredensborg, Olsen & Olsen, p. 133–137.

Nance, R.D., Miller, B.V., Keppie, J.D., Murphy, J.B., and Dostal, J., 2006, Acatlán Complex, southern Mexico: Record spanning the assembly and breakup of Pangea: Geology, v. 34, p. 857–860, doi: 10.1130/G22642.1.

Natal'in, B.A., 2007, Tectonic evolution of Mongolia, northeastern China, and southeastern Russia in the Paleozoic and early Mesozoic, in The Third International Workshop and Field Excursion for IGCP/Project 480, Beijing (China): Paris, International Geoscience Program, p. 1–3.

Natal'in, B.A., and Şengör, A.M.C., 2005, Late Palaeozoic to Triassic evolution of the Turan and Scythian platforms: The pre-history of the Palaeo-Tethyan closure: Tectonophysics, v. 404, p. 175–202, doi: 10.1016/j.tecto.2005.04.011.

Natal'in, B.A., Amato, J.M., Toro, J., and Wright, J.E., 1999, Paleozoic rocks of northern Chukotka Peninsula, Russian Far East: Implications for the tectonics of the Arctic region: Tectonics, v. 18, p. 977–1003, doi: 10.1029/1999TC900044.

Neumayr, M., 1885, Die geographische Verbreitung der Juraformation: Denkschriften der kaiserlichen Akademie der Wissenschaften (Wien), mathematisch-naturwissenscaftliche Classe, v. 50, p. 57–86.

Neumayr, M., 1887, Erdgeschichte, v. 2: Leipzig, Verlag des Bibliographischen Instituts, 880 p.

Newell, N.D., 1967, Revolutions in the History of Life, *in* Uniformity and Simplicity—A Symposium on the Principle of the Uniformity of Nature: Boulder, Geological Society of America Special Paper 89, p. 63–91.

Nie, S., Rowley, D.B., and Ziegler, A.M., 1990, Constraints on the locations of Asian microcontinents in the Paleo-Tethys during the Late Paleozoic, *in* McKerrow, W.S., and Scotese, C.R., eds., Paleozoic Paleogeography and Biogeography: Geological Society, London, Memoir 12, p. 397–409.

Nie, S.Y., An, Y., Rowley, D.B., and Jin, Y.G., 1994, Exhumation of the Dabie Shan ultra-high-pressure rocks and accumulation of the Songpan-Ganzi flysch sequence, central China: Geology, v. 22, p. 999–1002, doi: 10.1130/0091-7613(1994)022<0999:EOTDSU>2.3.CO;2.

Noble, P., and Renne, P., 1990, Paleoenvironmental and biostratigraphic significance of siliceous microfossils of the Permo-Triassic Redding section, Eastern Klamath Mountains, California: Marine Micropaleontology, v. 15, p. 379–391, doi: 10.1016/0377-8398(90)90021-D.

Noé, S.U., 1988, The Permian-Triassic boundary in the Southern Alps—a study of foraminiferal evolution: Berichte der Geologischen Bundesanstalt, Wien, v. 15, p. 19.

Norin, E., 1946, Geological Explorations in Western Tibet: The Sino-Swedish Expedition Publication 29, III: Geology, v. 7, 214 p. + 14 plates.

Norin, E., 1976, The "black slates" formations in the Pamirs, Karakoram and western Tibet: Atti dei Convegni Lincei 21 Colloquio Internazionale sulla Geotettonica delle Zone Orogeniche del Kashmir Himalaya Karakorum-Hindu Kush-Pamir: Roma, Accademia Nazionale dei Lincei, p. 245–264.

Norin, E., 1979, The relationships between the Tibetan platform and the Tarim basin: Bulletin of the Geological Institutions of the University of Uppsala, new series, v. 8, p. 17–34 + 2 plates.

Oaie, G., 1986, Sedimentology of the pebbly red sand-stone facies association within the Carapelit Formation, North Dobrogea: Revue Romaine de Géologie Géophysique et Géographie, v. 30 (Géologie), p. 59–70.

O'Brien, G.W., Milnes, A.R., Veeh, H.H., Heggie, D.T., Riggs, S.R., Cullen, D.J., Marshall, J.F., and Cook, P.J., 1990, Sedimentation dynamics and redox iron-cycling: controlling factors for the apatite-glauconite association on the East-Australian margin, *in* Notholt, A.J.G., and Jarvis, I., eds., Phosphorite Research and Development: Geological Society, London, Special Publication 52, p. 61–86.

Officer, C.B., and Page, J., 1996, The Great Dinosaur Extinction Controversy: Reading, Addison-Wesley, 209 p.

Officer, C.B., Hallam, A., Drake, C.L., and Devine, J.D., 1987, Late Cretaceous and paroxysmal Cretaceous/Tertiary extinctions: Nature, v. 326, p. 143–149, doi: 10.1038/326143a0.

Okada, H., and Kenyon-Smith, A.J., 2005, The Evolution of Clastic Sedimentology: Edinburgh, Dunedin Academic Press, 251 p.

Okay, A.İ., and Özgül, N., 1984, HP/LT metamorphism and the strucutre of the Alanya Massif, southern Turkey: an allochthonous composite tectonic sheet, *in* Dixon, J.E., and Robertson, A.H.F., eds., The Geological Evolution of the Eastern Mediterranean: Geological Society, London, Special Publication 17, p. 429–439.

Okay, A.İ., Şengör, A.M.C., and Shu, S.T., 1989, Coesite from the Dabie Shan eclogites, central China: European Journal of Mineralogy, v. 1, p. 595–598.

Okay, A.I., Şengör, A.M.C., and Satır, M., 1993, Tectonics of an ultra-high pressure metamorphic terrane: The Dabie Shan/Tongbai Shan orogen, China: Tectonics, v. 12, p. 1320–1334, doi: 10.1029/93TC01544.

Okay, A., Şengör, A.M.C., and Görür, N., 1994, The Black Sea: Kinematic history of its opening and its effect on the surrounding regions: Geology, v. 22, p. 267–270, doi: 10.1130/0091-7613(1994)022<0267:KHOTOO>2.3.CO;2.

Ota, A., and Isozaki, Y., 2006, Fusuline biotic turnover across the Guadalupian-Lopingian (Middle-Upper Permian) boundary in mid-oceanic carbonate buildups: Biostratigraphy of accreted limestone in Japan: Journal of Asian Earth Sciences, v. 26, p. 353–368, doi: 10.1016/j.jseaes.2005.04.001.

Ota, A., Kanmera, K., and Isozaki, Y., 2000, Stratigraphy of the Permian Iwato and Mitai formations in the Kamura area, southwest Japan: Maokouan, Wuchiapingian, and Changhsingian carbonates formed on paleo-seamount: The Journal of the Geological Society of Japan, v. 106, p. 853–864.

Owen, R. (Sir), 1845, Description of certain fossil crania discovered by A.G. Bain, Esq., in the sandstone rocks at the southeastern extremity of Africa, referable to different species of an extinct genus of Reptilia (*Dicynodon*), and indicative of a new tribe or suborder of Sauria: Transactions of the Geological Society of London, v. 2, no. 7, p. 233–240.

Owen, R. (Sir), 1860, Palæontology or a Systematic Summary of Extinct Animals and Their Geological Relations: Edinburgh, Adam and Charles Black, 420 p.

Özgül, N., 1976, Torosların bazı temel jeoloji özellikleri: Türkiye Jeoloji Kurumu Bülteni, v. 19, p. 65–78.

Paquier, J., 1910, Qu'est-ce que le Quiétisme? Questions Théologiques: Paris, Bloud et Cie, 122 p.

Parrish, J.T., and Ziegler, A.M., 1983, Upwelling in the Paleozoic Era, *in* Thiede, J., and Suess, E., eds., Coastal Upwelling: Its Sediment Record: New York, Plenum Press, v. B, p. 553–578.

Pasini, M., 1985, Biostratigrafia con i Foraminiferi del limite Formazioni a Bellerophon/Formazioni di Werfen fra Recaro e la Val Badia (Alpi Meridonali): Rivista Italiana Paleontologia e Stratigrafia, v. 90, p. 481–510.

Patterson, C., and Smith, A.B., 1987, Is the periodicity of extinctions a taxonomic artefact?: Nature, v. 330, p. 248–251, doi: 10.1038/330248a0.

Payne, J.L., Lehrmann, D.J., Follett, D., Seibel, M., Kump, L.R., Riccardi, A., Altıner, D., Sano, H., and Wei, J., 2007, Erosional truncation of uppermost Permian shallow-marine carbonates and implications for Permian-Triassic boundary events: Geological Society of America Bulletin, v. 119, p. 771–784, doi: 10.1130/B26091.1.

Pearce, J.A., and Mei, H., 1988, Volcanic rocks of the 1985 Tibet geotraverse: Lhasa and Golmud: Philosophical Transactions of the Royal Society of London, Series A, Mathematical and Physical Sciences, v. 327, p. 169–201, doi: 10.1098/rsta.1988.0125.

Peate, D.W., 1997, The Paraná-Etendeka province, *in* Mahoney, J.J., and Coffin, M.F., eds., Large Igneous Provinces—Continental, Oceanic, and Planetary Flood Volcanism: Washington, D.C., American Geophysical Union, Geophysical Monograph 100, p. 217–245.

Peryt, T.M., and Peryt, D., 1977, Zechstein foraminifera from the Fore-Sudetic monocline area (west Poland) and their palaeoecology: Annales de Société Géologique de Polonie, v. 47, p. 301–326.

Peters, J.A., 1949, Extinction: its causes and results: Biologist (Columbus, Ohio), v. 32, p. 3–8.

Phillips, J., 1840, Palaeozoic series, *in* Penny Cyclopædia: London, Charles Knight & Co., v. 17, p. 153–154.

Phillips, J., 1860, Life on Earth—Its Origin and Succession: Cambridge, Macmillan and Co., 224 p. + 1 frontispiece.

Piveteau, J., 1923, Sur un fragment de crâne de Dicynodon recuilli par H. Counillon dans les environs de Luang-Prabang (Haut-Laos): Bulletin du Service Géologique de l'Indochine, v. 12, fascicule 2, p. 1–7.

Plafker, G., and Berg, H.C., 1994, Overview of the geology and tectonic evolution of Alaska, *in* Plafker, G., and Berg, H.C., eds., The Geology of Alaska: Boulder, Geological Society of America, Geology of North America, v. G-1, p. 989–1021.

Poole, F.G., Perry, W.J., Jr., Madrid, R.J., and Amaya-Martinez, R., 2005, Tectonic synthesis of the Ouachita-Marathon-Sonora orogenic margin of southern Laurentia: Stratigraphic and structural implications for timing of deformational events and plate tectonic model, *in* Anderson, T.H., Nourse, J.A., McKee, J.W., and Steiner, M.B., eds., The Mojave-Sonora Megashear Hypothesis: Development, Assessment, and Alternatives: Boulder, Colorado, Geological Society of America Special Paper 393, p. 543–596, doi: 10.1130/2005.2393(21).

Popov, S.V., Rögl, F., Rozanov, A.Y., Steininger, F.F., Shcherba, I.G., and Kovac, M., eds., 2004, Lithological-Paleogeographic Maps of Paratethys—10 Maps Late Eocene-to Pliocene: Courier Forschungsinstitut Senckenberg 250, p. 1–46 + 10 maps.

Posenato, R., Pelikán, P., and Hips, K., 2005, Bivalves and brachiopods near the Permian-Triassic boundary from the Bükk Mountains (Bálvány-North Section, northern Hungary): Rivista Italiana di Paleontologia e stratigrafia, v. 111, p. 215–232.

Potter, P.E., Maynard, J.B., and Pryor, W.A., 1980, Sedimentology of Shale—Study Guide and Reference Source: New York, Springer-Verlag, 306 p.

Powers, C., and Bottjer, D.J., 2007, Bryozoan paleoecology indicates mid-Phanerozoic extinctions were the product of long-term environmental stress: Geology, v. 35, p. 995–998, doi: 10.1130/G23858A.1.

Price-Lloyd, N., and Twitchett, R.J., 2002, The Lilliput effect in the aftermath of the end-Permian extinction event, *in* Palaeontological Association 46th Annual Meeting, Department of Earth Sciences, University of Cambridge, 15–18 December 2002, Annual Conference Abstracts.

Prothero, D.R., 1994, The Eocene-Oligocene Transition—Paradise Lost: Critical Moments in Paleobiology and Earth History Series: New York, Columbia University Press, 291 p.

Ptolemy, C., 1482, Claudius Ptolemaeus Cosmographia: Ulm, Lienhart Holle (Leonhard Holl), unnumbered folios and tabulae.

Purdy, E.G., 2007, Comparison of taxonomic diversity, strontium isotope and sea-level patterns: International Journal of Earth Sciences, v. 97, no. 3, p. 651–664, doi: 10.1007/s00531-007-0177-z.

Ramovš, A., 1986, Reefbuilding organisms and reefs in the Permian of Slovenia, NW Yugoslavia: Memorie della Societa Geologica Italiana, v. 34, p. 189–193.

Randon, C., Wonganan, N., and Caridroit, M., 2007, Conodonts and radiolarians from northern Thailand: palaeogeographical outputs (Abstract), *in* 1er Symposium International de Paléobiogéographie Paris, 10–13 Juillet 2007, Université Pierre et Marie Curie (Paris 6): Paris, Muséum National d'Histoire Naturelle, CNRS, p. 84.

Rao, D.R., and Rai, H., 2007, Permian komatiites and associated basalts from the marine sediments of Chhongtash Formation, southeast Karakoram, Ladakh, India: Mineralogy and Petrology, v. 91, p. 171–189, doi: 10.1007/s00710-007-0206-4.

Raup, D.M., 1986, The Nemesis Affair—A Story of the Death of Dinosaurs and the Ways of Science: New York, W.W. Norton & Company, 220 p.

Raup, D.M., 1991, A kill curve for Phanerozoic marine species: Paleobiology, v. 17, p. 37–48.

Raup, D.M., and Sepkoski, J.J., Jr., 1982, Mass extinction in the marine fossil record: Science, v. 215, p. 1501–1503, doi: 10.1126/science.215.4539.1501.

Raymond, A., Parker, W.C., and Parrish, J.T., 1985, Phytogeography and paleoclimate of the early Carboniferous, *in* Tiffney, B.H., ed., Geological Factors and the Evolution of Plants: New Haven, Yale University Press, p. 169–222.

Razdan, M.L., Dhir, N.K., and Kacker, A.K., 1985, Record of late Permian ammonoid *Cyclolobus* from Zoji-La, Kargil District, Jammu and Kashmir: Current Science, v. 54, p. 187–188.

Reeburgh, W.S., Ward, B.B., Whalen, S.C., Sandbeck, K.A., Kilpatrick, K.A., and Kerkhof, L.J., 1991, Black Sea methane geochemistry: Deep-Sea Research, v. 38, supplement 2, p. 1189–1210.

Reeckmann, S.A., and Mebberson, A.J., 1984, Igneous intrusions in the northwest Canning Basin and their impact on oil exploration, *in* Purcell, P.G., ed., The Canning Basin, W.A., Proceedings of the Canning Basin Symposium: Perth, The Geological Society of Australia Incorporated and Petroleum Exploration Society of Australia Limited, p. 389–399.

Renne, P.R., Ernesto, M., Pacca, I.G., Coe, R.S., Glen, J.M., Prévot, M., and Perrin, M., 1992, The age of Paraná flood volcanism, rifting of Gondwanaland, and the Jurassic-Cretaceous boundary: Science, v. 258, p. 975–979, doi: 10.1126/science.258.5084.975.

Répelin, J., 1923, Sur un fragment de crâne de *Dicynodon* recueilli par H. Counillon dans les environs de Luang Prabang (Haut-Laos): Bulletin du Service Géologique de l'Indochine, v. 12, fascicule 2, p. 5–7.

Retallack, G.J., Veevers, J.J., and Morante, R., 1996, Global coal gap between Permian-Triassic extinction and Middle Triassic recovery of peat-forming plants: Geological Society of America Bulletin, v. 108, p. 195–207, doi: 10.1130/0016-7606(1996)108<0195:GCGBPT>2.3.CO;2.

Riccardi, A.L., Arthur, M.A., Kump, L.R., and D'Hondt, S., 2006, Sulfur isotopic evidence for chemocline upwater excursions during the end-Permian: Geochimica et Cosmochimica Acta v. 70, p. 5740–5752.

Riccardi, A.L., Kump, L.R., Arthur, M.A., and D'Hondt, S., 2007, Carbon isotopic evidence for chemocline upward excursions during the end-Permian event: Palaeogeography, Palaeoclimatology, Paleoecology v. 248, p. 73–81.

Ricci Lucchi, F., and Veggani, A., 1967, I Calcari a Lucina della formazione marnoso-arenacea Romagnola—nota preliminare: Giornale di Geologia (già Giornale di Geologia Practica): Annali del Museo Geologico di Bologna, Serie 2a, v. 34, fascicolo 1, p. 1–11.

Richter, B., and Fuller, M., 1996, Palaeomagnetism of the Sibumasu and Indochina blocks: implications for the extrusion tectonic model, *in* Hall, R., and Blundell, D., eds., Tectonic Evolution of Southeast Asia: Geological Society, London, Special Publication 106, p. 203–224.

Richter-Bernburg, G., 1955, Stratigraphische Gliederung des deutschen Zechsteins: Zeitschrift der Deutschen Geologischen Gesellschaft, v. 105, p. 593–645.

Richter-Bernburg, G., 1960, Zeitmessung geologischer Vorgänge nach Warven-Korrelationen im Zechstein: Geologische Rundschau, v. 49, p. 132–148, doi: 10.1007/BF01802401.

Ricou, L.-E., 1994, Tethys reconstructed: plates, continental fragments and their boundaries since 260 Ma from Central America to Southeastern Asia: Geodinamica Acta, v. 7, p. 169–218.

Ricou, L.-E., 1996, The plate tectonic history of the past Tethyan Ocean, *in* Nairn, A.E.M., Ricou, L.-E., Vrielynck, B., and Dercourt, J., eds., The Ocean Basins and Margins, The Tethys Ocean: New York, Plenum Press, v. 8, p. 3–70.

Rigby, J.F., and Shah, S.C., 1998, 'Gondwanaland' or 'Gondwana': Journal of African Earth Sciences, v. 27, supplement 1, p. iv.

Ritts, B.D., and Biffi, U., 2001, Mesozoic northeast Qaidam basin: Response to contractional reactivation of the Qilian Shan, and implications for the extent of Mesozoic intracontinental deformation in central Asia, *in* Hendrix, M.S., and Davis, G.A., eds., Paleozoic and Mesozoic Tectonic Evolution of Central and Eastern Asia: From Continental Assembly to Intracontinental Deformation: Boulder, Geological Society of America Memoir 194, p. 293–316.

Roecker, S.W., 1982, Velocity structure of the Pamir-Hindu Kush region: possible evidence of subducted crust: Journal of Geophysical Research, v. 87, p. 945–959, doi: 10.1029/JB087iB02p00945.

Rosendahl, W., Kempe, S., and Döppes, D., 2005, The scientific discovery of "Ursus spelaeus", *in* 11 Internationales Höhlenbär-Symposium, 29 September–2 October 2005, Naturhistorische Gesellschaft Nürnberg e.V.: Abhandlungen, v. 45, p. 199–214.

Rosenmüller, J.C., 1794, Quaedam de ossibus fossilibus animalis cujusdam, historiam ejus et cognitionem accuratiorem illustrantia, disertatio, quam d. 22 Octob. 1794. Ad disputandum proposuit Ioannes Christ. Rosenmüller Heßberga-Francus,LL. AA.M. in Theatro aanatomico Lipsiensi Prosector asumto socio Io. Christ. Heinroth Lips. Med. Stud. Cum tabula aenea, Leipzig, 34 p.

Rosenmüller, J.C., 1795, Beiträge zur Geschichte und näheren Kenntnis fossiler Knochen: Leipzig, Georg Emil Beer, 92 p.

Ross, C.A., and Ross, J.R.P., 1995, Foraminiferal zonation of the late Paleozoic depositional sequences: Marine Micropaleontology, v. 26, p. 469–478, doi: 10.1016/0377-8398(95)00010-0.

Ross, J.R.P., and Ross, C.A., 1991, Late Palaeozoic bryozoan biogeography, *in* McKerrow, W.S., and Scotese, C.R., eds., Palaeozoic Palaeogeography and Biogeography: The Geological Society, London, Memoir 12, p. 353–362.

Rossi, P., Guillot, S., and Menot, R.P., 2008, La branche varisque méridionale: Géochronique, no. 105, p. 45–49.

Rottura, A., Bargossi, G.M., Caggianelli, A., Del Moro, A., Visonà, D., and Trann, C.A., 1998, Origin and significance of the Permian high-K calc-alkaline magmatism in the central-eastern Southern Alps, Italy: Lithos, v. 45, p. 329–348, doi: 10.1016/S0024-4937(98)00038-3.

Rowley, D.B., Raymond, A., Parrish, J.T., Lottes, A.L., Scotese, C.R., and Ziegler, A.M., 1985, Carboniferous paleogeographic, phytogeographic and paleoclimatic reconstructions: International Journal of Coal Geology, v. 5, p. 7–42, doi: 10.1016/0166-5162(85)90009-6.

Ryan, W.B.F., 2009, Decoding the Mediterranean salinity crisis, *in* Bernoulli, D., Cita, M.B., and McKenzie, J.A., eds., Major Discoveries in Sedimentary Geology in the Mediterranean Realm from a Historical Perspective to New Developments: International Association of Sedimentology Special Publication (in press).

Ryan, W.B.F., and Cita, M.B., 1977, Ignorance concerning episodes of ocean-wide stagnation: Marine Geology, v. 23, p. 197–215, doi: 10.1016/0025-3227(77)90089-5.

Ryskin, G., 2003, Methane-driven oceanic eruptions and mass extinctions: Geology, v. 31, p. 741–744, doi: 10.1130/G19518.1.

Rzhevsky, Y.S., and Khramov, A.N., 1985, Palinspastic restorations for the South Tien Shan Mountains and Tajik depression based on paleomagnetic data: Journal of Geodynamics, v. 2, p. 119–126, doi: 10.1016/0264-3707(85)90005-5.

Salvador, A., ed., 1994, International Stratigraphic Guide—A Guide to Stratigraphic Classfication, Terminology, and Procedure, second edition: Trondheim, Norway, The International Union of Geological Sciences and, Boulder, Geological Society of America, 214 p.

Sanchez Carretero, R., Eguiluz, E., Pascual, E., and Carracedo, M., 1990, Igneous rocks, *in* Dallmeyer, R.D., and Martinez Garcia, E., eds., Pre-Mesozoic Geology of Iberia: Berlin, Springer-Verlag, p. 292–313.

Sarjeant, W.A.S., 1992, Müller, D.W., McKenzie, J.A., and Weissert, H., eds., 1991, *Modern Controversies in Geology. Evolution of Geological Theories in Sedimentology, Earth History and Tectonics* (book review): Geological Magazine, v. 129, p. 373.

Schaeffer, B., 1973, Fishes and the Permian-Triassic boundary, *in* Logan, A., and Hills, L.V., eds., The Permian and Triassic systems and their mutual boundary: Canadian Society of Petroleum Geologists Memoir 2, p. 493–497.

Schandelmeier, H., Reynods, P.-O., and Semtner, A.-K., 1997, Palaeogeographic-Palaeotectonic Atlas of North-Eastern Africa, Arabia, and

Adjacent Areas—Late Neoproterozoic to Holocene: Rotterdam, A.A. Balkema, 160 p. + 17 plates.

Schindewolf, O., 1950, Der Zeitfaktor in Geologie und Paläontologie—Akademische Antrittsvorlesung am 8. Juli 1948 im Auditorium Maximum der Universität Tübingen: Stuttgart, E. Schweizerbart'sche Verlagsbuchhandlung (Erwin Nägele), 114 p.

Schindewolf, O., 1954, Über die möglichen Ursachen der großen erdgeschichtlichen Faunenschnitte: Neues Jahrbuch für Geologie und Paläontologie. Monatshefte, v. 10, p. 457–465.

Schindewolf, O., 1962, Neokatastrophismus?: Zeitschrift der Deutschen Geologischen Gesellschaft, v. 114, p. 430–445.

Schmid, H.P., Harzhauser, M., Kroh, A., Coric, S., Rögl, F., and Schultz, O., 2001, Hypoxic events on a Middle Miocene carbonate platform of the Central Paratethys (Austria, Badenian, 14 Ma): Annalen des Naturhistorischen Museums in Wien, v. 102A, p. 1–50.

Scholten, R., 1974, Role of the Bosporus in Black Sea chemistry and sedimentation, *in* Degens, E.T., and Ross, D.A., eds., The Black Sea—Geology, Chemistry, and Biology: Tulsa, Oklahoma, American Association of Petroleum Geologists Memoir 20, p. 115–126.

Schubert, C.J., Durisch-Kaiser, E., Klauser, L., Vazquez, F., Wehrli, B., Holzner, C.P., Kipfer, R., Schmale, O., Greinert, J., and Kuypers, M.M.M., 2006, Recent studies on sources and sinks of methane in the Black Sea, *in* Neretin, L.N., ed., Past and Present Water Column Anoxia: Dordrecht, Springer, p. 419–441.

Scotese, C.R., 2001, Atlas of Earth History, Paleogeography: Arlington, Texas, PALEOMAP Project, v. 1, 52 p.

Searle, M.P., 1983, Stratigraphy, structure and evolution of the Tibetan-Tethys zone in Zanskar and Indus suture zone in the Ladakh Himalaya: Transactions of the Royal Society of Edinburgh, Earth Sciences, v. 73, p. 205–219.

Sedlock, R.L., Ortega-Gutiérrez, F., and Speed, R.C., 1993, Tectonostratigraphic Terranes and Tectonic Evolution of Mexico: Boulder, Geological Society of America Special Paper 278, 153 p.

Seghedi, A., and Oaie, G., 1986, Formaţiunea de Carapelit (Dobrogea de Nord): Faciesuri şi structuri sedimentare, 4. Stratigrafie: Ministerul Geologiei Institutul de Geologie şi Geofizică, Dări de Seama ale şedinţelor, v. 70–71, p. 19–37, 1 map and plates II–X.

Seilacher, A., 1999, Erdgeschichte als Langzeit-Experiment: Die grossen Revolutionen in der Entwicklung des Lebens: Eclogae Geologicae Helvetiae, v. 92, p. 73–79.

Self, S., Thordarson, T., and Keszthelyi, L., 1997, Emplacement of continental flood basalt lava flows, *in* Mahoney, J.J., and Coffin, M.F., eds., Large Igneous Provinces—Continental, Oceanic, and Planetary Flood Volcanism: American Geophysical Union, Geophysical Monograph 100, p. 381–410.

Şengör, A.M.C., 1979, Mid-Mesozoic closure of Permo-Triassic Tethys and its implications: Nature, v. 279, p. 590–593, doi: 10.1038/279590a0.

Şengör, A.M.C., 1982a, The classical theories of orogenesis, *in* Miyashiro, A., Aki, K., and Şengör, A.M.C., eds., Orogeny: Chichester, John Wiley & Sons, p. 1–48.

Şengör, A.M.C., 1982b, Eduard Suess' relations to the pre-1950 schools of thought in global tectonics: Geologische Rundschau, v. 71, p. 381–420, doi: 10.1007/BF01822372.

Şengör, A.M.C., 1983, Gondwana and "Gondwanaland": A discussion: Geologische Rundschau, v. 72, p. 397–400, doi: 10.1007/BF01765917.

Şengör, A.M.C., 1984, The Cimmeride orogenic system and the tectonics of Eurasia: Boulder, Geological Society of America Special Paper 195, 82 p.

Şengör, A.M.C., 1986, Die Alpiden und die Kimmeriden: die verdoppelte Geschichte der Tethys: Geologische Rundschau, v. 75, p. 501–510, doi: 10.1007/BF01820625.

Şengör, A.M.C., 1987, Tectonic subdivisions and evolution of Asia: Bulletin of the Technical University of İstanbul, v. 40 (Berker Festschrift), p. 355–435.

Şengör, A.M.C., 1990a, Tethys, Thethys, or Thetys? What, where, and when was it?: Comment: Geology, v. 18, p. 575, doi: 10.1130/0091-7613(1990)018<0575:CAROTT>2.3.CO;2.

Şengör, A.M.C., 1990b, Lithotectonic terranes and the plate tectonic theory of orogeny: a critique of the principles of terrane analysis, *in* Wiley, T.J., Howell, D.G., and Wong, F.L., eds., Terrane Analysis of China and the Pacific Rim: Houston, Circum-Pacific Council for Energy and Mineral Resources, Earth Science Series, v. 13, p. 9–44.

Şengör, A.M.C., 1990c, A new model for the late Palaeozoic-Mesozoic tectonic evolution of Iran and implications for Oman, *in* Robertson, A.H.F.,

Searle, M.P., and Ries, A.C., eds., Geological Society, London, Special Publication 49, p. 797–831.

Şengör, A.M.C., 1990d, Plate tectonics and orogenic research after 25 years: A Tethyan perspective: Earth-Science Reviews, v. 27, p. 1–201, doi: 10.1016/0012-8252(90)90002-D.

Şengör, A.M.C., 1991a, Difference between Gondwana and Gondwana-Land: Geology, v. 19, p. 287–288, doi: 10.1130/0091-7613(1991)019<0287:L>2.3.CO;2.

Şengör, A.M.C., 1991b, Orogenic architecture as a guide to size of ocean lost in collisional mountain belts: Bulletin of the Technical Univeristy of Istanbul, v. 44, p. 43–74.

Şengör, A.M.C., 1991c, Timing of orogenic events: a persistent geological controversy, *in* Müller, D.W., McKenzie, J.A., and Weissert, H., eds., Modern Controversies in Geology, Proceedings of the Hsü Symposium: London, Academic Press, p. 405–473.

Şengör, A.M.C., 1993, Some current problems on the tectonic evolution of the Mediterranean during the Cainozoic, *in* Boschi, E., Mantovani, E., and Morelli, A., eds., Recent Evolution and Seismicity of the Mediterranean Region, NATO ASI Series, Series C: Mathematical and Physical Sciences, v. 402: Dordrecht, Holland, Kluwer Academic Publishers, p. 1–51.

Şengör, A.M.C., 1994, Eduard Suess, *in* Eblen, R.A., and Eblen, W.R., eds., The Encyclopedia of the Environment: Boston, Houghton Mifflin Co., p. 676–677.

Şengör, A.M.C., 1998, Die Tethys: vor hundert Jahren und heute: Mitteilungen der Österreichischen Geologischen Gesellschaft, v. 89, p. 5–176.

Şengör, A.M.C., 2000, Die Bedeutung von Eduard Suess (1831–1914) für die Geschichte der Tektonik: Berichte der Geologischen Bundesanstalt, v. 51, p. 57–72.

Şengör, A.M.C., 2006, Grundzüge der geologischen Gedanken von Eduard Suess Teil 1: Einführung und erkenntnistheoretische Grundlagen: Jahrbuch der Geologischen Bundesanstalt, Wien, v. 146, p. 265–301.

Şengör, A.M.C., and Dewey, J.F., 1990, Terranology: Vice or Virtue?: Philosophical Transactions of the Royal Society of London, v. 331A, p. 457–477, doi: 10.1098/rsta.1990.0083.

Şengör, A.M.C., and Hsü, K.J., 1984, The Cimmerides of Eastern Asia: History of the eastern end of Palaeo-Tethys, *in* Buffetaut, E., Jaeger, J.-J., and Rage, J.-C., eds., Paléogéographie de l'Inde, du Tibet et du Sud-Est Asiatique: Confrontations des Données Paléontologiques aves les Modèles Géodynamiques, Mémoires de la Société Géologique de France, nouvelle série no. 147, p. 139–167.

Şengör, A.M.C., and Natal'in, B.A., 1996, Palaeotectonics of Asia: Fragments of a synthesis, *in* Yin, A., and Harrison, M., eds., The Tectonic Evolution of Asia, Rubey Colloquium: Cambridge, Cambridge University Press, p. 486–640.

Şengör, A.M.C., and Natal'in, B.A., 2001, Rifts of the world, *in* Ernst, R., and Buchan, K., eds., Mantle Plumes: Their Identification Through Time: Boulder, Geological Society of America Special Paper 352, p. 389–482.

Şengör, A.M.C., and Natal'in, B.A., 2004, Phanerozoic analogues of Archaean oceanic basement fragments: Altaid ophiolites and ophirags, *in* Kusky, T.M., ed., Precambrian Ophiolites and Related Rocks, Developments in Precambrian Geology: Amsterdam, Elsevier, v. 13, p. 675–726.

Şengör, A.M.C., and Yılmaz, Y., 1981, Tethyan evolution of Turkey: a plate tectonic approach: Tectonophysics, v. 75, p. 181–241, doi: 10.1016/0040-1951(81)90275-4.

Şengör, A.M.C., Altıner, D., Cin, A., Ustaömer, T., and Hsü, K.J., 1988, Origin and assembly of the Tethyside orogenic collage at the expense of Gondwana-Land, *in* Audley-Charles, M.G., and Hallam, A., eds., Gondwana and Tethys: Geological Society, London, Special Publication 37, p. 119–181.

Şengör, A.M.C., Cin, A., Rowley, D.B., and Nie, S.Y., 1991, Magmatic evolution of the Tethysides: a guide to reconstruction of collage history: Palaeogeography, Palaeoclimatology, Palaeoecology, v. 87, p. 411–440, doi: 10.1016/0031-0182(91)90143-F.

Şengör, A.M.C., Natal'in, B.A., and Burtman, V.S., 1993a, Evolution of the Altaid tectonic collage and Palaeozoic crustal growth in Eurasia: Nature, v. 364, p. 299–307, doi: 10.1038/364299a0.

Şengör, A.M.C., Cin, A., Rowley, D.B., and Nie, S.Y., 1993b, Space-time patterns of magmatism along the Tethysides: a preliminary study: The Journal of Geology, v. 101, p. 51–84.

Şengör, A.M.C., Yılmaz, Y., and Sungurlu, O., 1984, Tectonics of the Mediterranean Cimmerides: nature and evolution of the western termination of Palaeo-Tethys, *in* Dixon, J.E., and Robertson, A.H.F., eds., Geological Evolution of the Eastern Mediterranean: Geological Society, London, Special Publication 17, p. 77–112.

Şengör, A.M.C., Niemi, N., McQuarrie, N., Friedrich, A.M., and Wernicke, B., 2001, A large, subduction-generated syntaxis in the Gondwanides of southern Gondwana-Land: Geological Society of America Abstracts with Programs, v. 33, no. 6, p. A-397.

Şengör, A.M.C., Atayman, S., and Özeren, S., 2008, A scale of greatness and causal classification of mass extinctions: implications for mechanisms: Proceedings of the National Academy of Sciences of the United States of America, v. 105, p. 13,736–13,740, doi: 10.1073/pnas.0805482105.

Sepkoski, J.J., Jr., 1981, A factor analytic description of the marine fossil record: Paleobiology, v. 7, p. 36–53.

Sepkoski, J.J., Jr., 1986, Phanerozoic overview of mass extinctions, in Raup, D.M., and Jablonski, D., eds., Patterns and Processes in the History of Life: Berlin, Springer-Verlag, p. 277–295.

Sepkoski, J.J., Jr., 1989, Periodicity in extinction and the problem of catastrophism in the history of life: Journal of the Geological Society, v. 146, p. 7–19, doi: 10.1144/gsjgs.146.1.0007.

Sepkoski, J.J., Jr., 1990, The taxonomic structure of periodic extinction, in Sharpton, V.I., and Ward, P.D., eds., Global Catastrophies in Earth History: Boulder, Geological Society of America Special Paper 247, p. 33–44.

Sereno, P.C., 1999, The evolution of dinosaurs: Science, v. 284, p. 2137–2147, doi: 10.1126/science.284.5423.2137.

Sheldon, N., 2006, Abrupt chemical weathering increase across the Permian-Triassic boundary: Palaeogeography, Palaeoclimatology, Palaeoecology, v. 231, p. 315–321, doi: 10.1016/j.palaeo.2005.09.001.

Shen, S.Z., Archbold, N.M., and Shi, G.R., 2001, A Lopingian (late Permian) brachiopod fauna from the Qubuerga Formation at Shemgmi in the Mount Qomolangma region of southern Xizang (Tibet), China: Journal of Paleontology, v. 75, p. 274–283, doi: 10.1666/0022-3360(2001)075<0274: ALLPBF>2.0.CO;2.

Shen, S.Z., Mu, L., and Zakharov, Y.D., 2004, Roadoceras (Permian Ammonoidea) from the Qubuerga Formation in the Mt. Everest area in southern Tibet: Gondwana Research, v. 7, p. 863–869, doi: 10.1016/S1342-937X(05)71070-4.

Shen, S.Z., Cao, C., Henderson, C.M., Wang, X., Shi, G.R., Wang, Y., and Wang, W., 2006, End-Permian mass extinction pattern in the northern peri-Gondwanan region: Palaeoworld, v. 15, p. 3–30, doi: 10.1016/j.palwor.2006.03.005.

Sheng, J.Z., Chen, C.Z., Wang, Y.G., Rui, L., Liao, Z.T., Yuji, B., Ishii, K.I., Nakazawa, K., and Nakamura, K., 1984, Permian-Triassic boundary in middle and eastern Tethys: Journal of the Faculty of Science of Hokkaido University, series IV, v. 21, p. 133–181.

Sheng, J.-Z., Rui, L., and Chen, C.Z., 1985, Permian and Triassic sedimentary facies and paleogeography of south China, in Nakazawa, K., and Dickins, J.M., eds., The Tethys—Her Paleogeography and Paleobiogeography from Paleozoic to Mesozoic: Tokyo, Tokai University Press, p. 59–80.

Shi, G.R., 2006, The marine Permian of east and northeast Asia: an overview of biostratigraphy, palaeobiogeography and palaeogeographical implications: Journal of Asian Earth Sciences, v. 26, p. 175–206, doi: 10.1016/j.jseaes.2005.11.004.

Shishkhin, M.A., 1997, Post-extinction events in the land tetrapod communities during Permo-Triassic faunal turnover (abstract): http://www.gli.cas.cz/_abstr/0000003b.htm.

Sidor, C.A., O'Keefe, F.R., Damiani, R., Steyer, J.S., Smith, R.M.H., Larsson, H.C.E., Sereno, P.C., Ide, O., and Maga, A., 2005, Permian tetrapods from the Sahara show climate-controlled endemism in Pangaea: Nature, v. 434, p. 886–889, doi: 10.1038/nature03393.

Signor, P.W., III, and Lipps, J.H., 1982, Sampling bias, gradual extinction patterns, and catastrophes in the fossil record, in Silver, L.T., and Schultz, P.H., eds., Geological Implications of Impacts of Large Asteroids and Comets on the Earth: Boulder, Geological Society of America Special Publication 190, p. 291–296.

Sigogneau, D., 1970, Révision Systématique des Gorgonopsiens Sud-Africains: Paris, Cahiers de Paléontologie, Éditions du Centre National de la Recherche Scientifique, 416 p. + 93 plates.

Sinha-Roy, S., and Furnes, H., 1978, Geochemistry and geotectonic implications of basic volcanic rocks in the lower Gondwana sequence (Upper Palaeozoic) of the Sikkim Himalayas: Geological Magazine, v. 115, p. 427–436.

Sinha-Roy, S., and Furnes, H., 1980, Alkaline vs calc-alkaline continental magmatism: implications for Gondwanic rift in the Himalayas: Geological Magazine, v. 117, p. 624–629.

Sjostrom, D.J., Hendrix, M.S., Badamgarav, D., Graham, S.A., and Nelson, B.K., 2001, Sedimentology and provenance of Mesozoic nonmarine strata in western Mongolia: A record of intracontinental deformation, in Hendrix, M.S., and Davis, G.A., eds., Paleozoic and Mesozoic Tectonic Evolution of Central and Eastern Asia: From Continental Assembly to Intracontinental Deformation: Boulder, Geological Society of America Memoir 194, p. 361–388.

Slipper, I.J., 2005, Ostracod diversity and sea-level changes in the Late Cretaceous of southern England: Palaeogeography, Palaeoclimatology, Palaeoecology, v. 225, p. 266–282, doi: 10.1016/j.palaeo.2005.06.014.

Slomp, C.P., and van Capellan, P., 2007, The global marine phosphorous cycle: sensitivity to oceanic circulation: Biogeosciences, v. 4, p. 155–171.

Smith, A.G., 1971, Alpine deformation and the oceanic areas of the Tethys, Mediterranean and Atlantic: Geological Society of America Bulletin, v. 82, p. 2039–2070, doi: 10.1130/0016-7606(1971)82[2039:ADATOA]2.0.CO;2.

Sobel, E.R., Arnaud, N., Jolivet, M., Ritts, B.D., and Brunel, M., 2001, Jurassic to Cenozoic exhumation history of the altyn Tagh range, northwest China, constrained by $^{40}Ar/^{39}Ar$ and apatite fission track thermochronology, in Hendrix, M.S., and Davis, G.A., eds., Paleozoic and Mesozoic Tectonic Evolution of Central and Eastern Asia: From Continental Assembly to Intracontinental Deformation: Boulder, Geological Society of America Memoir 194, p. 247–267.

Sokołowski, S., ed., 1970, Geology of Poland, Stratigraphy, part 1 Pre-Cambrian and Palaeozoic: Warsaw, Publishing House Wyadawnictwa Geologiczne, v. 1, 651 p. + 1 map.

Solignac, M., and Berkaloff, E., 1934, Le Permien Marin de l'extrême-sud tunisien: 1. Considérations générales: le Djebel Tebaga: Régence de Tunis—Protectorat Français Direction des Travaux Publics—Service des Mines, Mémoires du Service de la Carte Géologique de Tunisie, nouvelle série, no. 1, p. 3–72 + 1 plate.

Sougy, J., 1962, West African foldbelt: Geological Society of America Bulletin, v. 73, p. 871–876, doi: 10.1130/0016-7606(1962)73[871:WAFB]2.0.CO;2.

Sperling, E.A., and Ingle, J.C., Jr., 2006, A Permian-Triassic boundary section at Quinn River Crossing, northwestern Nevada, and implications for the cause of the Early Triassic chert gap on the western Pangean margin: Geological Society of America Bulletin, v. 118, p. 733–746, doi: 10.1130/B25803.1.

Stampfli, G.M., 1978, Étude Géologique Générale de l'Elburz Oriental au S de Gonbad-e-Qabus Iran NE [Ph.D. thèse]: Genève, l'Université de Genève, no. 1868, 329 p. + 2 foldouts.

Stampfli, G.M., and Borel, G.D., 2002, A plate tectonic model for the Paleozoic and Mesozoic constrained by dynamic plate boundaries and restored synthetic oceanic isochrons: Earth and Planetary Science Letters, v. 196, no. 1, p. 17–33, doi: 10.1016/S0012-821X(01)00588-X.

Stampfli, G., Marcoux, J., and Baud, A., 1991, Tethyan margins in space and time: Palaeogeography, Palaeoclimatology, Palaeoecology, v. 87, p. 373–409, doi: 10.1016/0031-0182(91)90142-E.

Stampfli, G.M., Mosar, J., Favre, P., Pillevuit, A., and Vannay, J.-C., 2001, Permo-Mesozoic evolution of the western Tethys realm: the Neo-Tethys East Mediterranean basin connection, in Ziegler, P.A.,, Cavazza, W., Robertson, A.H.F., and Crasquin-Soleau, S., eds., Peri-Tethys Rift/Wrench Basin and Passive Margins: Mémoires du Muséum National d'Histoire Naturelle, v. 186, p. 51–108.

Stanley, S.M., 1987, Extinction: New York, Scientific American Library, 242 p.

Stanley, S.M., and Yang, X., 1994, A double mass extinction at the end of the Paleozoic era: Science, v. 266, p. 1340–1344, doi: 10.1126/science.266.5189.1340.

Stehli, F.G., 1973, Permian brachiopods, in Hallam, A., ed., Atlas of Palaeobiogeography: Amsterdam, Elsevier, p. 142–149.

Steiner, M.B., Eshet, Y., Rampino, M.R., and Schwindt, D.M., 2003, Fungal abundance spike and the Permian-Triassic boundary in the Karoo Supergroup (South Africa): Palaeogeography, Palaeoclimatology, Palaeoecology, v. 194, p. 405–414, doi: 10.1016/S0031-0182(03)00230-X.

Stevens, C.H., Stone, P., and Miller, J.S., 2005, A new reconstruction of the Paleozoic continental margin of southwestern North America: Implications for the nature and timing of continental truncation and the possible role of the Mojave-Sonora megashear, in Anderson, T.H., Nourse, J.A., McKee, J.W., and Steiner, M.B., eds., The Mojave-Sonora Megashear Hypothesis: Development, Assessment, and Alternatives: Boulder, Colorado, Geological Society of America Special Paper 393, p. 597–618, doi: 10.1130/2005/2393(22).

Stevens, H.N., 1908, Ptolemy's Geography. A Brief Account of all the Printed Editions Down to 1730: Henry Stevens, Son and Stiles, 62 p. (there is an undated reprint by Theatrvm Orbis Terrarvm, Ltd., Amsterdam).

Steyer, J.S., Damiaini, R., Sidor, C.A., O'Keefe, R., Larsson, H.C.E., Maga, A., and Ide, O., 2006, The vertebrate fauna of the Upper Permian of Niger. IV. *Nigerpeton ricqlesi* (Temnospondyli: Cochleosauridae), and the edopoid colonization of Gondwana: Journal of Vertebrate Paleontology, v. 26, p. 18–28, doi: 10.1671/0272-4634(2006)26[18:TVFOTU]2.0.CO;2.

Steyer, S., 2009, The geological and palaeontological exploration of Laos—following in the footsteps of J.B.H. Counillon and A. Pavie: Journal of the Geological Society (in press).

Stille, H., 1924, Grundfragen der Vergleichenden Tektonik: Berlin, Gebrüder Borntraeger, 443 p.

Stille, H., 1929, Tektonische Formen in Mitteleuropa und Mittelasien: Zeitschrift der Deutschen Geologischen Gesellschaft, v. 81, p. 2–9.

Stille, H., 1940, Einführung in den Bau Amerikas: Berlin, Gebrüder Borntraeger, 717 p.

Stow, D.A.W., and Lovell, J.P.B., 1979, Contourites: Their recognition in modern and ancient sediments: Earth-Science Reviews, v. 14, p. 251–291, doi: 10.1016/0012-8252(79)90002-3.

Strauss, H., 2006, Anoxia through time, *in* Neretin, L.N., ed., Past and Present Water Column Anoxia: Dordrecht, Springer-Verlag, p. 3–19.

Suess, E., 1860, Über die Wohnsitze der Brachiopoden—II. Abschnitt. Die Wohnsitze der fossilen Brachiopoden: Sitzungsberichte der kaiserlichen Akademie der Wissenschaften, mathematisch-naturwissenschaftliche Classe, no. 39, p. 151–206.

Suess, E., 1862, untitled letter, *in* Bericht vom 31 Juli 1862, Verhandlungen der kaiserlich-königlichen geologischen Reichsanstalt, Jahrbuch der königlich-kaiserlichen geologischen Reichsanstalt, v. 12, p. 258.

Suess, E., 1875, Die Entstehung der Alpen: Wien, W. Braumüller, 168 p.

Suess, E., 1885, Das Antlitz der Erde, v. Ib: Leipzig, F. Tempsky, Prag and G. Freytag, p. 311–778.

Suess, E., 1888, Das Antlitz der Erde, v. II: Leipzig, F. Tempsky, Prag and Wien, and G. Freytag, 704 p.

Suess, E., 1893, Are great ocean depths permanent? Natural Science, v. 2, p. 180–187.

Suess, E., 1901, Das Antlitz der Erde, Bd. III1 (Dritter Band. Erste Hälfte): Leipzig, F. Tempsky, Prag and Wien, and G. Freytag, 508 p.

Suess, E., 1904, Farewell lecture by Professor Edward Suess on resigning his professorship translated by Charles Schuchert: The Journal of Geology, v. 12, p. 264–275.

Suess, E., 1909, Das Antlitz der Erde, v. III2 (Dritter Band. Zweite Hälfte. Schluss des Gesamtwerkes): Leipzig, F. Tempsky, Wien and G. Freytag, 789 p.

Sun, K., 2006, The Cathaysia flora and the mixed Late Permian Cathaysian-Angaran floras in East Asia: Journal of Integrative Plant Biology, v. 48, p. 381–389, doi: 10.1111/j.1744-7909.2006.00207.x.

Szabo, F., and Kheradpir, A., 1978, Permian and Triassic stratigraphy, Zagros basin, south-west Iran: Journal of Petroleum Geology, v. 1, p. 57–82, doi: 10.1111/j.1747-5457.1978.tb00611.x.

Tabor, N.J., Montañez, I.P., Steiner, M.B., and Schwindt, D., 2007, δ^{13}C values of carbonate nodules across the Permian-Triassic boundary in the Karoo Supergroup (South Africa) reflect a stinking sulfurous swamp, not atmospheric CO_2: Palaeogeography, Palaeoclimatology, Palaeoecology, v. 252, p. 370–381, doi: 10.1016/j.palaeo.2006.11.047.

Takemura, A., Aita, Y., Hori, R.S., Higuchi, Y., Spörli, K.B., Campbell, H.J., Kodama, K., and Sakai, T., 2002, Triassic radiolarians from the ocean-floor sequence of the Waipapa terrane at Arrow Rocks, Northland, New Zealand: New Zealand Journal of Geology and Geophysics, v. 45, p. 289–296.

Takin, M., 1972, Iranian geology and continental drift in the Middle East: Nature, v. 235, p. 147–150, doi: 10.1038/235147a0.

Tankard, A.J., Jackson, M.P.A., Eriksson, K.A., Hobday, D.K., Hunter, D.R., and Minter, W.E.L., 1982, Crustal Evolution of Southern Africa—3.8 Billion Years of Earth History: New York, Springer-Verlag, 523 p.

Tappan, H., and Loeblich, A.R., Jr., 1988, Foraminiferal evolution, diversification, and extinction: Journal of Paleontology, v. 62, p. 695–714.

Taylor, P.T., 2004, Extinctions in the History of Life: Cambridge, Cambridge University Press, 191 p.

Tazawa, J., and Chen, Z.Q., 2005, Middle Permian brachiopods from the Tumenling Formation in the Wuchang area, southern Heilongjiang, NE China, and their palaeobiogeographical implications: Journal of Asian Earth Sciences, v. 26, p. 327–338, doi: 10.1016/j.jseaes.2005.06.008.

Teagle, D.A.H., and Wilson, D.S., 2007, Leg 206 synthesis: Initiation of drilling an intact section of upper oceanic crust formed at a superfast spreading rate at site 156 in the eastern equatorial Pacific, *in* Teagle, D.A.H., Wilson, D.S., Acton,

G.D., and Vanko, D.A., eds., Proceedings of the Ocean Drilling Program, scientific results, v. 206, p. 1–15, doi: 10.2973/odp.proc.sr.206.001.2007

Teilhard de Chardin, P., and Licent, E., 1924, Observations géologiques sur la bordure occidentale et méridionale de l'Ordos: Bulletin de la Société Géologique de France, série 4, v. 24, p. 49–91.

Thanh, T.-D., Janvier, P., and Phuong, T.H., 1996, Fish suggests continental connections between the Indochina and South China blocks in Middle Devonian time: Geology, v. 24, p. 571–574, doi: 10.1130/0091-7613(1996) 024<0571:FSCCBT>2.3.CO;2.

Thompson, S.C., Weldon, R.J., Rubin, C.M., Abdrakhmatov, K., Molnar, P., and Berger, G.W., 2002, Late Quaternary slip rates across the central Tien Shan, Kyrgyzstan, central Asia: Journal of Geophysical Research, v. 107, p. ETG 7-1–ETG 7-32, no. B9, 2203, doi: 10.1029/2001JB000596.

Timmerman, M.J., 2004, Timing, geodynamic setting and character of Permo-Carboniferous magmatism in the foreland of the Variscan orogen, NW Europe, *in* Wilson, M., Neumann, E.-R., Davies, G.R., Timmerman, M.J., Heeremans, M., and Larsen, B.T., eds., Permo-Carboniferous Magmatism and Rifting in Europe: Geological Society, London, Special Publication 223, p. 41–74.

Tissot, B., Deroo, G., and Herbin, J.P., 1979, Oceanic matter in Cretaceous sediments of the north Atlantic Ocean: contributions to sedimentology and paleogeography, *in* Talwani, M., Hay, W.W., and Ryan, W.B.F., eds., Deep Drilling Results in the Atlantic Ocean: Continental Margins and Paleoenvironments: Washington, D.C., American Geophysical Union, Maurice Ewing Series, v. 3, p. 362–374.

Tong, J., 1993, Biotic mass extinction and biotic alteration at the Permo-Triassic boundary, Foraminifera, *in* Yang, Z., Wu, S., Yin. H., Xu, G., Zhang, K., and Bi, X., eds., Permo-Triassic events of south China: Bejing, Geological Publishing House, p. 90–97.

Topuz, G., Altherr, R., Satır, M., and Schwarz, W.H., 2004, Lowgrade metamorphic rocks from the Pulur complex, NE Turkey: implications for the pre-Liassic evolution of the eastern Pontides: International Journal of Earth Sciences, v. 93, p. 72–91, doi: 10.1007/s00531-003-0372-5.

Tozer, E.T., 1989, Tethys, Thethys, or Thetys? What, where, and when was it?: Geology, v. 17, p. 882–884, doi: 10.1130/0091-7613(1989)017<0882: TTTOTW>2.3.CO;2.

Tozer, E.T., 1990, Tethys, Thethys, or Thetys? What, where, and when was it?: Reply: Geology, v. 18, p. 575–576, doi: 10.1130/0091-7613(1990)018 <0575:CAROTT>2.3.CO;2.

Trappe, J., 1994, Pangean phosphorites—ordinary phosphorite genesis in an extraordinary world, *in* Embry, A.F., Beauchamp, B., and Glass, D.J., eds., Pangea: Global Environments and Resources: Alberta, Canadian Society of Petroleum Geologists Memoir 17, p. 469–478.

Trümpy, R., 1960, Über die Perm-Trias-Grenze in Ostgrönland und über die Problematik stratigraphischer Grenzen: Geologische Rundschau, v. 49, p. 97–103, doi: 10.1007/BF01802396.

Trunkó, L., 1996, Geology of Hungary: Beiträge zur Regionalen Geologie der Erde: Berlin, Gebrüder Borntraeger, 464 p.

Twitchett, R.J., 2007, Oxygen as a control on the Permian-Triassic evolution of the marine biosphere: Geological Society of America Abstracts with Programs, v. 39, no. 6, p. 25.

Vallaux, C., 1933, Géographie Générale des Mers: Paris, Félix Alcan, 795 p.

Van der Voo, R., 1993, Paleomagnetism of the Atlantic, Tethys and Iapetus Oceans: New York, Cambridge University Press, 411 p.

Vannay, J.C., 1993, Geologie des chaines du Haut-Himalaya et du Pir Panjal au Haut-Lahul (NW Himalaya, Inde); paleogeographie et tectonique [Ph.D. thesis]: Lausanne, Universite de Lausanne, Memoires de Geologie, no. 16, 148 p.

van Wees, J.-D., Stephenson, R.S., Ziegler, P.A., Bayer, U., McCann, T., Dadlez, R., Gaupp, R., Narkiewicz, M., Bitzer, F., and Scheck, M., 2000, Origin of the southern Permian basin, central Europe: Marine and Petroleum Geology, v. 17, p. 43–59, doi: 10.1016/S0264-8172(99)00052-5.

Veevers, J.J., ed., 2000, Billion-Year Earth History of Australia and Neighbours in Gondwanaland: Sydney, Gemoc Press, 388 p.

Veevers, J.J., 2001, Atlas of Billion-Year Earth History of Australia and Neighbours in Gondwanaland: Sydney, Gemoc Press, 76 p.

Veevers, J.J., 2004, Gondwanaland from 650–500 Ma assembly through 320 Ma merger in Pangea to 185–100 Ma breakup: supercontinental tectonics via stratigraphy and radiometric dating: Earth-Science Reviews, v. 68, p. 1–132, doi: 10.1016/j.earscirev.2004.05.002.

Veevers, J.J., and Tewari, R.C., 1995, Gondwana Master Basin of Peninsular India—Between Tethys and the Interior of the Gondwanaland Province of Pangea: Boulder, Geological Society of America Memoir 187, 72 p.

Villeneuve, M., 2008, Review of the orogenic belts on the western side of the West African craton: the Bassarides, Rokelides and Mauritanides, *in* Ennih, N., and Liégeois, J.-P., eds., The Boundaries of the West African Craton: Geological Society, London, Special Publication 297, p. 169–201, doi: 10.1144/SP297.8.

Vincent, S.J., and Allen, M.B., 2001, Sedimentary record of Mesozoic intracontinental deformation in the eastren Junggar Basin, northwest China: Response to orogeny at the Asian margin, *in* Hendrix, M.S., and Davis, G.A., eds., Paleozoic and Mesozoic Tectonic Evolution of Central and Eastern Asia: From Continental Assembly to Intracontinental Deformation: Boulder, Geological Society of America Memoir 194, p. 341–360.

Vinogradov, P., 1958, A. Beleuli (Série de...), *in* Likharev, B.K., ed., U.R.S.S., Lexique Stratigraphique International: Paris, Centre National de la Recherce Scientifique, v. 2, fascicule 1, p. 188.

Virtasalo, J.J., Kohonen, T., Vuorinen, I., and Huttula, T., 2005, Sea bottom anoxia in the Archipelago Sea, northern Baltic Sea—implications for phosphorus remineralization at the sediment surface: Marine Geology, v. 224, p. 103–122, doi: 10.1016/j.margeo.2005.07.010.

Visscher, H., Brinkhuis, H., Dilcher, D.L., Elsik, W.C., Eshet, Y., Looy, C.V., Rampino, M.R., and Traverse, A., 1996, The terminal Paleozoic fungal event: Evidence of terrestrial ecosystem destabilization and collapse: Proceedings of the National Academy of Sciences of the United States of America, v. 93, p. 2155–2158, doi: 10.1073/pnas.93.5.2155.

Visscher, H., Kerp, H., Clement-Westerhof, J.A., and Looy, C.V., 2001, Permian floras of the Southern Alps: Annali dei Museo Civico di Scienze Naturali di Brescia Monografiea 25, p. 117–123.

Waagen, W., 1891, Salt Range Fossils. Geological Results: Palæontologia Indica, Being Figures and Descriptions of the Organic Remains Produced During the Progress of the Geological Survey of India, ser. XIII, v. 4, no. 2, p. 89–242.

Wahnschaffe, F., and Schucht, F., 1921, Geologie und Oberflächengestaltung des Norddeutschen Flachlandes: Stuttgart, J. Engelhorns Nachfolger, 472 p. + 1 map.

Wang, H.Z., chief compiler, 1985, Atlas of the Paleogeography of China: Beijing, Cartographic Publishing House, 301 p.

Ward, P., 1994, The End of Evolution—On Mass Extinctions and the Preservation of Biodiversity: New York, Bantam Books, 302 p.

Ward, P., 2000, Rivers in Time—Earth's Mass Extinctions: New York, Columbia University Press, 315 p. (updated and entirely revised version of 1994).

Ward, P.D., 2006, Dinosaurs, Birds, and Earth's Ancient Atmosphere—Out of Thin Air: Washington, D.C., Joseph Henry Press, 282 p.

Ward, P.D., and Berner, R.A., 2007, Oxygen, carbon dioxide, and mass extinctions: Boulder, Geological Society of America Abstracts with Programs, v. 39, no. 6, p. 24.

Ward, P.D., Botha, J., Buick, R., De Kock, M.O., Erwin, D.H., Garrison, G., Kirschvink, J., and Smith, R., 2005, Abrupt and gradual extinction among Late Permian land vertebrates in the Karoo basin, South Africa: Science, v. 308, doi: 10.1126/science.1107068.

Wartes, M.A., Carroll, A.R., and Greene, T.J., 2002, Permian sedimentary record of the Turpan-Hami basin and adjacent regions, northwest China: Constraints on post-amalgamation tectonic evolution: Geological Society of America Bulletin, v. 114, p. 131–152, doi: 10.1130/0016-7606(2002)114 <0131:PSROTT>2.0.CO;2.

Webby, B.D., 2004, Introduction, *in* Webby, B.D., Paris, F., Droser, M.L., and Percival, I.G., eds., The Great Ordovician Biodiversification Event: New York, Columbia University Press, p. 1–37.

Webby, B.D., Paris, F., Droser, M.L., and Percival, I.G., eds., 2004, The Great Ordovician Biodiversification Event: New York, Columbia University Press, 484 p.

Weissert, H., 1981, The environment of deposition of black shales in the early Cretaceous: an ongoing controversy, *in* Warme, J.E., Douglas, R.G., and Winterer, E.L., eds., The Deep Sea Drilling Project: A Decade of Progress: Tulsa, Oklahoma, Society of Economic Paleontologists and Mineralogists (SEPM) Special Publication 32, p. 547–560.

Wendt, J., Hayer, J., and Bavandpur, A.R., 1997, Stratigraphy and depositional environment of Devonian sediments in northeast and east-central Iran: Neues Jahrbuch für Geologie und Palaontologie: Abhandlungen, v. 206, p. 277–322.

Wendt, J., Kaufmann, B., Belka, Z., Farsan, N., and Bavandpur, A.R.K., 2002, Devonian/Carboniferous stratigraphy, facies patterns and palaeogeography of Iran. Part I. Southeastern Iran: Acta Geologica Polonica, v. 52, p. 129–168.

Wendt, J., Kaufmann, B., Belka, Z., Farsan, N., and Bavandpur, A.R.K., 2005, Devonian/Carboniferous stratigraphy, facies patterns and palaeogeography of Iran. Part II. Northern and central Iran: Acta Geologica Polonica, v. 55, p. 31–97.

Wignall, P., 1994, Black Shales: Oxford, Oxford University Press, Oxford Science Publications, Oxford Monographs on Geology and Geophysics, no. 30, 127 p.

Wignall, P.B., and Newton, R., 2003, Deep-water records from the upper Permian and lower Triassic of South Tibet and British Columbia Evidence for a diachronous mass extinction: Palaios, v. 18, p. 153–167, doi: 10.1669/0883-1351(2003)18<153:CDRFTU>2.0.CO;2.

Wignall, P.B., Morante, R., and Newton, R., 1998, The Permo-Triassic transition in Spitsbergen: $\delta^{13}C_{org}$ chemostratigraphy, Fe and S geochemistry, facies, fauna and trace fossils: Geological Magazine, v. 135, p. 47–62, doi: 10.1017/S0016756897008121.

Wignall, P.B., Newton, R., and Brookfield, M.E., 2005, Pyrite framboid evidence for oxygen-poor deposition during the Permian-Triassic crisis in Kashmir: Palaeogeography, Palaeoclimatology, Palaeoecology, v. 216, p. 183–188, doi: 10.1016/j.palaeo.2004.10.009.

Willis, B., 1932, Isthmian links: Geological Society of America Bulletin, v. 43, p. 917–952.

Willis, K.J., and McElwain, J.C., 2002, The Evolution of Plants: Oxford, Oxford University Press, 378 p.

Wilson, J.T., 1963, Hypothesis of Earth's behaviour: Nature, v. 198, p. 925–929, doi: 10.1038/198925a0.

Wilson, J.T., 1966, Did the Atlantic close and then re-open?: Nature, v. 211, p. 676–681, doi: 10.1038/211676a0.

Wilson, M., Neumann, E.-R., Davies, G.R., Timmerman, M.J., Heeremans, M., and Larsen, B.T., eds., 2004, Permo-Carboniferous Magmatism and Rifting in Europe: Geological Society, London, Special Publication 223, 498 p.

Winterer, E.L., 1976, Anomalies in the tectonic evolution of the Pacific, *in* Sutton, G.H., Manghnani, M.H., and Moberly, R., eds., The Geophysics of the Pacific Ocean Basin and Its Margin: Washington, D.C., American Geophysical Union, Geophysical Monograph 19, p. 269–278.

Wolfart, R., and Wittekindt, H., 1980, Geologie von Afghanistan: Berlin, Gebrüder Borntraeger, Beiträge zur Regionalen Geologie der Erde, v. 14, 500 p. + 2 maps.

Wonganan, N., and Caridroit, M., 2006, Middle to Upper Permian radiolarian faunas from chert blocks in Pai area, northwestern Thailand: Eclogae Geologicae Helvetiae, v. 99, p. 133–139, doi: 10.1007/s00015-006-0610-5.

Wopfner, H., 1984, Permian deposits of the Southern Alps as product of initial Alpidic taphrogenesis: International Journal of Earth Scie nces, v. 73, p. 259–277.

Wu, Y.S., 2006, Ecological selection in end-Permian mass extinction and its climatic implication: Earth & Life, v. 1, no. 1, p. 16–30.

Wu, Y.S., Yang, W., Jiang, H.X., and Fan, S.S., 2007, Temporal pattern of oceanic anoxia across the Permian-Triassic boundary: Earth & Life, v. 1, no. 2, p. 15–22.

Xiao, W.J., and He, H.Q., 2005, Early Mesozoic thrust tectonics of the northwest Zhejiang region (Southeast China): Geological Society of America Bulletin, v. 117, p. 945–961, doi: 10.1130/B25417.1.

Xiao, W.J., Windley, B.F., Hao, J., and Zhai, M.G., 2003, Accretion leading to collision and the Permian Solonker suture, Inner Mongolia, China: termination of the Central Asian orogenic belt: Tectonics, v. 22, p. 8-1–8-6, doi: 10.1029/2002TC001484.

Xiao, W.-J., Zhang, L.-C., Qin, K.Z., Sun, S., and Li, J.L., 2004, Paleozoic accretionary and collisional tectonics of the eastern Tianshan (China): Implications for the continental growth of central Asia: American Journal of Science, v. 304, p. 370–395, doi: 10.2475/ajs.304.4.370.

Yamamoto, S., Alcuaskas, J.B., and Crozier, T.E., 1976, Solubility of methane in distilled water and seawater: Journal of Chemical and Engineering Data, v. 21, p. 78–80.

Yang, F.Q., and Wang, H.M., 1999, Ammonoid recovery after end-Permian mass extinction in south China, *in* Proceeding of the International Conference on Pangea and the Paleozoic-Mesozoik Transition: Wuhan, China University of Geosciences Press, p. 64–68.

Yang, Z.Y., Wu, S.B., and Yang, Q., 1981, Permo-Triassic boundary of south China: Journal of the Wuhan College of Geology, v. 1, p. 4–15.

Yang, Z.Y., Chen, Y.Q., and Wang, H.Z., 1986, The Geology of China: Oxford, Clarendon Press, Oxford Monographs on Geology and Geophysics, 303 p.

Yanko-Hombach, V., Gilbert, A.S., Panin, N., and Dolukhanov, P.M., eds., 2007, The Black Sea Flood Question—Changes in Coastline, Climate and Human Settlement: Berlin, Springer, 971 p.

Yin, A., Nie, S., Craig, P., Harrison, T.M., Ryerson, F.J., Qian, X.L., and Yang, G., 1998, Late Cenozoic tectonic evolution of the southern Chinese Tian Shan: Tectonics, v. 17, p. 1–27, doi: 10.1029/97TC03140.

Yin, H., Wu, S., Din, M., Zhang, K., Tong, J., and Yang, F., 1994, The Meishan section candidate of the global stratotype section and point (GSSP) of the Permian–Triassic boundary (PTB): Albertiana, v. 14, p. 15–31.

Yin, H.F., Yang, F.Q., Zhang, K.X., and Yang, W.P., 1986, A proposal to the biostratigraphic criterion of Permian/Triassic boundary: Memorie della Società Geologica Italiana, v. 34, p. 329–344.

Yin, H.F., Zhang, K.X., Tong, J.N., Yang, Z.Y., and Wu, S.B., 2001, The global stratotype section and point (GSSP) of the Permian-Triassic boundary: Episodes, v. 24, p. 102–114.

Young, C.C., 1935, On two skeletons of Dicynodontia from Sinkiang: Bulletin of the Geological Society of China, v. 14, p. 483–517.

Young, C.C., 1939, Additional Dicynodontia from Sinkiang: Bulletin of the Geological Society of China, v. 19, p. 111–146.

Yuan, P.L., and Young, C.C., 1934, On the occurrence of Lystrosaurus in Sinkiang: Bulletin of the Geological Society of China, v. 13, p. 575–580.

Zanchi, A., and Gaetani, M., 1994, Introduction to the geological map of the north Karakorum terrain from the Chapursan valley to the Shimshal Pass 1:150,000 scale: Rivista Italiana di Paleontologia e Stratigrafia, v. 100, p. 125–136 + map.

Zhang, L.F., Ai, Y.L., Li, X.P., Rubatto, D., Song, B., Williams, S., Song, S.G., Ellis, D., and Liou, J.G., 2007, Triassic collision of western Tianshan orogenic belt, China: evidence from SHRIMP U-Pb dating of zircon from HP/UHP eclogitic rocks: Lithos, v. 96, p. 266–280, doi: 10.1016/j.lithos.2006.09.012.

Zhang, P., Burchfiel, B.C., Molnar, P., Zhang, W.Q., Jiao, D.C., Deng, Q.D., Wang, Y.P., Royden, L., and Song, F.M., 1990, Late Cenozoic tectonic evolution of the Ninxia-Hui autonomous region, China: Geological Society of America Bulletin, v. 102, p. 1484–1498, doi: 10.1130/0016-7606(1990)102<1484:LCTEOT>2.3.CO;2.

Zhang, Y.X., 1996, Dynamics of CO_2-driven lake eruptions: Nature, v. 379, p. 57–59, doi: 10.1038/379057a0.

Zhang, Y.X., and Kling, G.W., 2006, Dynamics of lake eruptions and possible ocean eruptions: Annual Review of Earth and Planetary Sciences, v. 34, p. 293–324, doi: 10.1146/annurev.earth.34.031405.125001.

Zharkov, M.A., 1984, Paleozoic Salt Bearing Formations of the World: Berlin, Springer-Verlag, 427 p.

Zheng, J.S., Mermet, J.-F., Toutin-Morin, N., Hanes, J., Gondono, A., Morin, R., and Féraud, G., 1991–1992, Datation ^{40}Ar-^{39}Ar du magmatisme et de filons minéralisés permiens en Provence orientale (France): Geodinamica Acta, v. 5, p. 203–215.

Zhou, D., and Graham, S.A., 1996, The Songpan-Ganzi complex of the west Qinling Shan as a Triassic remnant ocean basin, *in* Yin, A., and Harrison, M.B., eds., The Tectonic Evolution of Asia: Cambridge, Cambridge University Press, p. 281–299.

Ziegler, A.M., Hulver, M.L., and Rowley, D.B., 1997, Permian world topography and climate, *in* Martini, I.P., ed., Late Glacial and Postglacial Environmental Changes—Quaternary, Carboniferos-Permian, and Proterozoic: New York, Oxford University Press, p. 111–146 + 1 plate.

Ziegler, P.A., 1988, Evolution of the Arctic-north Atlantic and the western Tethys: Tulsa, Oklahoma, American Association of Petroleum Geologists Memoir 48, 198 p. + 30 plates.

Ziegler, P.A., 1989, Evolution of Laurussia: Dordrecht, Kluwer Academic Publishers, 102 p. + 14 plates.

Ziegler, P.A., 1990, Geological Atlas of Western and Central Europe 1990, second edition: Den Haag, Shell Internationale Petroleum Maatschappij B.V., 239 p. + 56 plates.

Zonenshain, L.P., Kuzmin, M.I., and Natapov, L.M., 1990. Geology of the USSR: A plate tectonic synthesis: AGU Geodynamic Series, v. 21, 242 p.

MANUSCRIPT ACCEPTED BY THE SOCIETY 9 AUGUST 2008

Index

Note: Page numbers with "f" and "n" indicate material in figures and footnotes, respectively.